AF816945

THE POLITICS OF LAND AND FOOD SCARCITY

In recent years the issue of food security has taken centre stage on the global agenda. Since the 2007/8 food price crisis, a number of works have been published on the topic, addressed from various perspectives: economic, social and cultural, environmental, agronomic and climate change. Very rarely is there a comprehensive approach, which also includes the crucial issue of politics.

Through a multidisciplinary approach, this book provides an overview of the new global challenges connected with land, food supply and agriculture. It also contributes to engagement in a new global food policy, through a political analysis of land and food scarcity, including 'land-grabs' by affluent countries in poorer nations. It does not simply raise the issues; rather it aspires to move forward the debate that has started with the G20 meetings. It discusses how national governments, local agricultural policies and supranational entities are facing the new scenario of feeding a growing population when land resources are limited and subject to competing claims.

Paolo De Castro is Professor of Agricultural Economics and Policy at the University of Bologna, Italy, and Chair of the Committee on Agriculture and Rural Development at the European Parliament.

Felice Adinolfi is Professor of Agricultural Economics and Policy at the University of Bologna, Italy.

Fabian Capitanio is Associate Professor of Agricultural Economics and Policy at the University of Naples, Italy.

Salvatore Di Falco is Lecturer in Environment and Development Economics at the London School of Economics, UK.

Angelo Di Mambro is a freelance journalist based in Brussels, Belgium.

Other books in the Earthscan Food and Agriculture Series

Food Systems Failure
The Global Food Crisis and the Future of Agriculture
Edited by Chris Rosin, Paul Stock and Hugh Campbell

Understanding the Common Agricultural Policy
By Berkeley Hill

The Sociology of Food and Agriculture
By Michael Carolan

Competition and Efficiency in International Food Supply Chains
Improving Food Security
By John Williams

Organic Agriculture for Sustainable Livelihoods
Edited by Niels Halberg and Adrian Muller

The Politics of Land and Food Scarcity
By Paolo De Castro, Felice Adinolfi, Fabian Capitanio, Salvatore Di Falco and Angelo Di Mambro

Principles of Sustainable Aquaculture
Promoting Social, Economic and Environmental Resilience
By Stuart Bunting

THE POLITICS OF LAND AND FOOD SCARCITY

Edited by Paolo De Castro
with Felice Adinolfi, Fabian Capitanio,
Salvatore Di Falco and Angelo Di Mambro

Routledge
Taylor & Francis Group
LONDON AND NEW YORK

earthscan
from Routledge

This first edition published 2013 by Routledge
2 Park Square, Milton Park, Abingdon, Oxon, OX14 4RN

Simultaneously published in the USA and Canada
by Routledge
711 Third Avenue, New York, NY 10017

Routledge is an imprint of the Taylor & Francis Group, an informa business

© 2013 Paolo De Castro, Felice Adinolfi, Fabian Capitanio, Salvatore Di Falco and Angelo Di Mambro

The right of Paolo De Castro, Felice Adinolfi, Fabian Capitanio, Salvatore Di Falco and Angelo Di Mambro to be identified as authors of this work has been asserted by them in accordance with sections 77 and 78 of the Copyright, Designs and Patents Act 1988.

All rights reserved. No part of this book may be reprinted or reproduced or utilised in any form or by any electronic, mechanical, or other means, now known or hereafter invented, including photocopying and recording, or in any information storage or retrieval system, without permission in writing from the publishers.

Trademark notice: Product or corporate names may be trademarks or registered trademarks, and are used only for identification and explanation without intent to infringe.

British Library Cataloguing in Publication Data
A catalogue record for this book is available from the British Library

Library of Congress Cataloging-in-Publication Data
The politics of land and food scarcity / Paolo De Castro... [et al.].
p. cm. – (Earthscan food and agriculture)
Includes bibliographical references and index.
1. Food supply. 2. Food security. I. De Castro, Paolo, 1958-
HD9000.5.P624 2012
338.1'9–dc23
2012017255

ISBN13: 978-0-415-63823-4 (hbk)
ISBN13: 978-0-415-63824-1 (pbk)
ISBN13: 978-0-203-08413-7 (ebk)

Typeset in Sabon by Saxon Graphics Ltd, Derby

CONTENTS

This book is the product of the joint effort of the following authors: Paolo De Castro, Felice Adinolfi, Fabian Capitanio, Salvatore Di Falco and Angelo Di Mambro.

Their specific collaborations are as follows:
Part I: Paolo De Castro, Felice Adinolfi, Fabian Capitanio, Angelo Di Mambro and Jorgelina Di Pasquale;
Part II: Paolo De Castro, Felice Adinolfi, Angelo Di Mambro and Fabian Capitanio;
Part III: Paolo De Castro, Felice Adinolfi and Salvatore Di Falco.

FOREWORD

The topic of food security, in other words the possibility of gaining access to food, has returned to centre stage with the recent boom in food prices. That is not to say the problem did not exist before. Hunger has continued, despite the technical progress recorded in recent decades, to torment hundreds of millions of people in the world's poorer areas. Great responsibility falls on the shoulders of wealthier nations, whose commitment in combating hunger in the world has been insufficient both in terms of resources and that of instruments. However, despite the weakness and inefficiency that has often been a hallmark of intervention in the international community, at the end of the 1990s and the start of the new millennium we were able to record some major results in tackling poverty in less developed areas of the planet. The results were also the fruit of a new assumption of responsibility: a possible target would be that of halving the percentage of the world's population in conditions of extreme poverty by 2015.

However, it must be acknowledged that neither the political will nor the necessary financial resources have materialised to reach this target. There has not been the great international mobilisation that would have been necessary.

Moreover, such ambitions had not allowed for entry into a completely new scenario. In the space of a few years many of the convictions that had been formed in the previous decades were swept away. Farm productivity which had made extraordinary progress in the decades of the Green Revolution is today growing at very low rates. Research in the agricultural sector is stagnating and does not offer concrete prospects for the return of a cycle of strong productivity increases. The effect of this evolution is now evident: after a long period – roughly since the end of World War II – of falling prices of agricultural commodities, the world seems to have entered what Paolo De Castro calls the era of scarcity. The increase in population, new dietary styles that are becoming increasingly widespread in the emerging areas of the world and the progressive slowdown in farm productivity have driven us to a new perspective which questions our planet's capacity to satisfy the substantial increase in food demand.

The first effects of this change are right in front of us. The aim of reducing the world population living below the poverty line seems to have disappeared for good and preoccupations are such that even more developed contexts are driven to take measures to prevent the occurrence of any difficulties on a national basis. On many occasions, such actions prove uncoordinated and counter-productive. An example is restrictive measures on exports, which have for some years become an integral part of commercial policies of the big producers of foodstuffs and have had great responsibility in making price rises more acute: attitudes that also give us the idea of an unprepared international community and that today, more than ever, need to build coordinated measures on a global level to tackle one of the great emergencies of our time.

This book tells us precisely what is happening, analyses in depth the motivations and implications of this new era of food scarcity and suggests the horizons within which to seek appropriate solutions. It has the merits not only of providing detail but also of being highly readable, studiously avoiding technical excesses and allowing the reader to absorb, as in a story, the extensive information embedded in the text.

The expertise and competence of Paolo De Castro, as well as that of his working group which accompanied him in this project, are a clearly visible hallmark which makes this book interesting for a far wider readership than just area specialists, for whom it will nonetheless represent an excellent point of reference, also thanks to the authoritative contributions to the text.

Romano Prodi

LIST OF ILLUSTRATIONS

Figures

Boxes

INTRODUCTION AND ACKNOWLEDGEMENTS

The idea of writing this book first came in May 2011 straight after a meeting organised by the European Parliament on the subject of the security of food supply. The guest of the initiative was Romano Prodi who, with a lucid well-reasoned analysis, partly the fruit of his recent experience as a committed politician and academic in China and Africa, drew a somewhat alarming picture of the international situation and its foreseeable development.

The European Parliament was already well aware that food security is not just a question of distributing resources among rich and poor countries, but goes well beyond this to become a structural issue of global importance. Yet Romano Prodi's speech lent such a sense of urgency that we were led to speed up the debate and promote concrete initiatives straight away, aware of the role and responsibilities falling on Europe's shoulders, still the world's largest importer and exporter of food products.

Thus, a few weeks later, a delegation from the Agriculture Committee of the European Parliament went, for the first time, to the USA to hold discussions with the political authorities in Washington and start up a dialogue on the subject between the two most important players in the world food trade. The aim was to share the reasons behind this great emergency and begin to identify concrete solutions to be translated both into proposals to be submitted to the international community and into agricultural policy choices.

The facts concerning the growth in demand and supply were, and continue to be, unambiguous: farm production is not keeping pace with demand. The increase in population and even more marked rise in consumption especially in several of the Earth's major areas, such as China, India, Brazil and Russia, are increasing food demand at alarming rates, while the availability of agricultural land and its productivity are declining. This is occurring at such a rate that a practice commonly called 'land-grabbing' has emerged: millions of hectares of land in Africa are being acquired, and not only by foreign investors.

This phenomenon, perhaps more so than others, gives us the idea of major changes under way, which are given extensive airing in this book: the new strategic value of land as a resource and the far-reaching changes affecting the structure of food markets.

With the contribution of the working group that has supported me in this matter, I have sought to explain, hopefully in a readable form, the detailed reasons for, and implications of, this new scenario into which we have been catapulted. I have tried to explain the import of an emergency which, in our latitudes where expenditure counts for less than 15 per cent of overall household expenditure, risks being viewed as remote, while it is actually dramatically close. We are experiencing a rapid expansion in the area of world poverty, and consumer growth is challenging to the extreme the capacity of the whole agri-food business to respond. In 2050 there will be more than nine billion inhabitants on Earth, about a third more than today, and to satisfy food demand we will need to increase agricultural production by 70 per cent compared with the present day. What is more, it will have to be done more sustainably than in the past.

Produce more, polluting less: an objective that is tough to achieve without a committed intervention by the international community and an equally decisive U-turn in public commitment to research and innovation, which has declined in recent years at alarming rates. In this framework we wish to suggest what avenues might, in our opinion, be pursued to tackle a situation that has assumed the proportions of an imminent global emergency.

The credit for this work, targeting both policy-makers and all those who wish to understand in greater depth the dimensions of this new scenario of food scarcity, should go to all those who have directly or indirectly collaborated. In particular, thanks are due to my co-authors, Felice Adinolfi (University of Bologna), Fabian Capitanio (University of Naples), Salvatore Di Falco (London School of Economics) and Angelo Di Mambro with whom I created this project and saw it through to its conclusion, whose passion and expertise characterised this work.

I wish to thank Policy Department B of the European Parliament, whose studies have been an important source for the animated discussions that led to the drafting of the book.

I cannot avoid thanking all those who accompany my daily work. I am referring to my historical assistant, Dr Antonella Alvisini, Dr Antonella Rossetti who diligently manages my work in Brussels, Bruno Calzia and Marco Barbetta; their contribution has been essential.

Romano Prodi deserves special thanks, not only for his introduction but especially for his decisive contribution in triggering the reflections contained in this book. I am likewise grateful to Franz Fischler, Michael Taylor, Martin Schulz and Joseph Daul who have, with their authoritative points of view, enriched the treatment of the topics handled, giving us essential assistance in interpreting a scenario in which changes are both rapid and complex. Comments and help were also gratefully received from Denis Pantini and Andrea Zaghi (Nomisma), and Geoffroy Enjolras (INRA). The original Italian version was translated into English by Mark Walters (Part I) and Paul Tout (Parts II and III).

Finally, I would like to thank all the staff who provide the daily back-up for my work at the European Parliament, whose industry and patience were essential to the success of this book.

I hope that this work can be really useful both to facilitate the understanding of the scenario in which we are called upon to reform what is historically the most important European policy, and to help implement a new generation of policy framework in agriculture and rural areas by effectively responding to the challenges of our time.

Paolo De Castro, European Parliament, Brussels, October 2011

PART I

Food insecurity

1

THERE ARE MORE OF US; WE'RE BETTER OFF. AND NOW?

The return of scarcity

The issue of access to, and availability of, food on a global basis, known as *food security*,[1] has been making headway on the political and media agenda since the start of the twenty-first century. What now gives cause for concern is the increase in food prices. From 1945, there was a long period of stagnating and declining prices on agricultural markets, punctuated only by a spurt in the 1970s, in concomitance with the oil shock that followed the embargo decided by the OPEC[2] countries. Since the mid 1990s there has been an inverse trend, with sharp peaks in farm commodity prices in 2007/2008 and 2010/2011. The most recent forecasts indicate a sizeable rise in prices for the coming years. Despite the debate on the role played by the various factors involved, most academics agree that the long era of abundant food, available at low prices, is over, and has given way to an era of new scarcity.

The phenomenon is usually linked to projections on demographic growth, according to which, in 2050, there will be more than nine billion inhabitants on the planet. Too many to be fed. It is the return of one of the recurrent nightmares of classical economics. As early as 1798, the economist and clergyman Thomas Malthus with his *An Essay on the Principle of Population as it Affects the Future Improvement of Society* maintained that society would reach what he called 'breakpoints', i.e. a temporary halt to economic development. These would be repeated cyclically, insofar as the population increase accompanying development would lead people to cultivate marginal, fairly unproductive land, with consequent inability to satisfy the food demand of a population that would grow more rapidly than food availability. This was a work of outstanding importance in modern economic thought, which led to the birth of a real doctrine, Malthusianism, based on the principle of limited natural resources: a theory which, with the passing of the years, has lost importance since the contribution of technological progress has fortunately allowed us to increase agricultural yields at rates which would have been unthinkable at the time when Malthus published his essay.

Aside from Malthus, in tackling the issue of security in the food supply the analysis of population dynamics has given way to examination of distribution dynamics. The problem, as posed in recent years, is not scarcity, but the inequitable distribution of food resources, which has ended up penalising vast populous areas in the world. Yet the problem has not been solved. Quite the opposite, it has recently been aggravated, leading the number of undernourished people worldwide to exceed one billion. New, complex variables have entered the stage: these have brought back the prophecies formulated by the Anglican clergyman at the end of the eighteenth century, and encouraged some to give them a modern twist. There has arisen a sort of neo-Malthusian movement which, in its various hues, views with pessimism the recent evolution of the relations between food supply and demand, maintaining that world production will, in future, be increasingly less able to keep up with the increase in demand, leading to progressive widening of the areas and populations threatened by hunger.

Despite not belonging to the ranks of neo-Malthusians, we share the concerns of those who feel the renewed topicality of the nexus between economic and demographic growth as one of the main factors of change in a new scenario, viewed by many as food scarcity. Though inter-related with the broader issue of natural resource depletion, food scarcity assumes connotations of greater or equal urgency compared with other very delicate problems, such as climate change, conveyed more loudly by public opinion.

However we believe, as we will seek to explain in the course of this book, that the objective of making supply meet demand growth is not impossible, but can be achieved with the contribution of research and innovation as well as suitable policies to be constructed at the global level. At the same time, we are aware that the issue of food supply is very delicate, often triggers knee-jerk reactions that cannot cover exhaustively the quantity and quality of the factors involved, but are also conscious that attributing the origin of complex scenarios to single causes is often only the first step to failure to understand them completely. Thus, in seeking a single culprit, some blame the food price rise exclusively on biofuels, others on never well-identified speculators, others again on 'industrial' agriculture or on trade liberalisation. Regrettably, the *reductio ad unum* is a very widespread attitude when speaking of the food issue: an attitude that we will studiously seek to avoid. Witch-hunts have never really helped to understand reality.

Economic theory tells us that the rise in prices is due to a change in market equilibrium, caused by the increasing gap between the demand and supply of a certain product. In the case of agricultural markets, there are concerns regarding, chiefly, the increase in food demand, which in recent years has advanced at sustained rates, especially in the world's so-called emerging areas. In countries like India, China and Brazil, per capita wealth, and hence spending capacity, have increased appreciably. This means not only that food demand will rise, fuelled by the population increase, but also that it will change its connotations radically, with diets and eating habits becoming richer and more complex.

From the land to the city

According to FAO (Food and Agriculture Organisation of the United Nations) estimates, world population will exceed nine billion in 2050. This represents an increase of about one-third against the current population of 6.9 billion, numbers that come as no surprise if we look at population trends in the last 100 years. Population increases forecast for the coming decades will in fact be lower than in the past: the population increase of over 30 per cent predicted by the FAO for the next 40 years is well below the relative growth in the past four decades, during which the population more than doubled. However, it is the number in absolute terms which gives cause for concern. The planet will be inhabited by about 2.5 billion more people in the space of a few decades: in less than 20 years the Earth will have a billion more inhabitants and mouths to feed.

The largest increases will take place in developing countries, while the population in high-income economies will remain almost stable and in some areas, especially in some regions of Europe, there may even be population declines. By contrast, in Africa the population is expected to double, growing from one to two million by 2050. In emerging areas, growth will continue to be sustained, especially in India, while China's growth should slow down; absolute increases will remain appreciable. In these two countries, which now make up over one-third of the world's population, the number of inhabitants is expected to rise from the current 2.5 billion to 3.2 by the year 2050.

To summarise, the key contributor of global population increase in the years to come will be the planet's poorer areas and emerging countries. Without a similar increase in food production, the former areas risk a considerable increase in those with inadequate access to food. This possibility would appear to compromise in the short term the first of what have been called the *millennium development goals* (Box 1.1), in other words goals of sustainable development and the war against hunger which the countries belonging to the UN have undertaken to reach by the year 2015. For the emerging countries, the implications are more complex, with repercussions

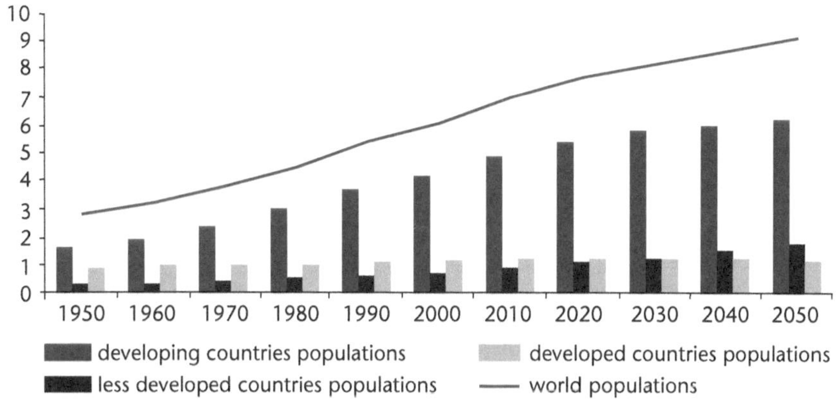

FIGURE 1.1 World population growth (in billions)
Source: UN

on a global scale: the growth dynamics of emerging economies represents one of the greatest threats to the equilibrium of markets in the short term and, in the medium–long term, cast doubts on certainties as regards the security of food supply, established in wealthier countries, keeping pace with processes of industrial and post-industrial growth. The global effects of population increases in these areas will be strictly linked to migration intensity between country and town, which will accompany economic growth and the rise in incomes. This has already been observed in the history of the development of so-called advanced economies.

Around 50 per cent of the world's population is now distributed in urban settlements, the other half in rural areas. In 1950 only 28 per cent of the population lived in large urban agglomerations; in 2050 this percentage is destined to reach 70 per cent. Compared with today, there will be 19 more cities with over ten million inhabitants and five of these will be in Asia where the trend toward metropolitan concentration will be particularly marked.

This phenomenon will affect the population in rural areas, such that the FAO estimates the population reduction active in agriculture in the next 40 years at around 30 per cent. However, it will also contribute to orienting consumer choices for much of the world's population towards products with larger contents of services (starting from processing) and thus bring them closer to the food styles in the planet's wealthier areas.

The most important contribution to the standardisation of diets will be made by the expansion of the middle classes in emerging areas. Individual income in countries like India, Brazil and China rose at sustained rates in recent years, only to slow down, but not stop, during this long phase of world economic recession. The cases that stand out most are those of China and India which have recorded annual growth rates close to double figures in the years immediately prior to the recession and which are forecast, according to the International Monetary Fund, to continue their trend at least for the next 20 years.

To understand the effects that this new wealth will produce on food demand let us call upon another economist, Ernst Engel, whose law, as it is still taught in university lecture halls, explains the relation between income changes and expenditure on consumption. As early as 1857, the German economist observed that as income rose, the share of income allocated to food consumption decreased, as with the share to satisfy the whole

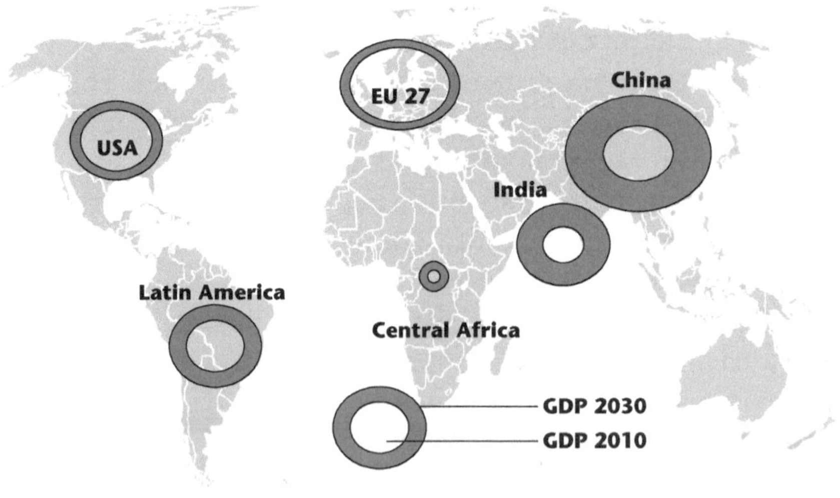

FIGURE 1.2 Gross domestic product (GDP) in some areas of the world. 2010–2030 growth projections
Source: IMF

range of the other primary needs (clothing, washing etc.), in favour of goods termed 'superior', used to meet needs that go beyond 'prime necessity', such as transport and free time.

This means two things: the first is that as incomes grow, expenditure on food consumption does not grow in proportion; the second is that food habits change, producing what economists call the 'substitution effect'. In practice, within the basket of household food expenditure, some products are replaced with others considered of greater value and quality. As populations gradually become richer, in their diets the unprocessed starch products (like rice and flour) are replaced by products with a higher protein content (such as meat, milk and other dairy products) and by processed products with greater value added, promoting a process of dietary convergence worldwide along the models of richer populations.

This trend will involve several billion people in emerging countries. In China, individual meat demand per annum is forecast to increase by over 28 kg in the next 40 years (over 50 per cent more than current demand), while average consumption of all emerging and developing countries is expected to rise by over 6 kg per capita. The same trend is forecast for dairy products which will see an increase in demand of about 70 per cent against current levels.[3]

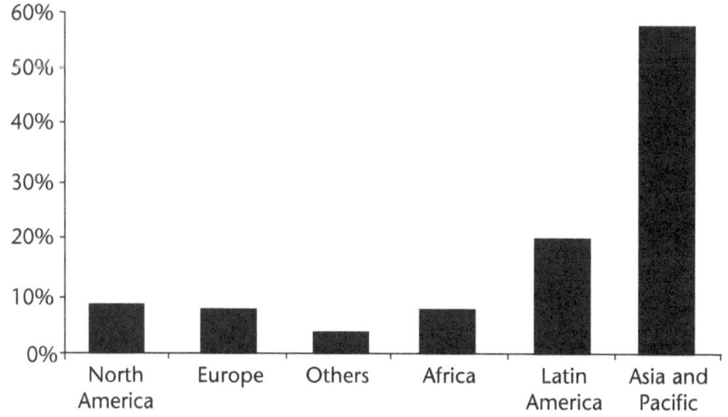

FIGURE 1.3 Meat consumption forecasts: 2010–2020 projections
Source: OECD and FAO

All this will have a multiplying effect also on the demand for some agricultural raw materials, like soya and wheat, which form the basis of animal feed. Albeit with wide variability dependent on livestock systems in use, between two and four kilos of grain are required to produce one kilo of chicken, while to produce a beef steak of the same weight between seven and ten kilos of grain are required. A medium protein diet now accounts for roughly four times the contribution of cereals present in a vegetarian diet.

How does all this impact upon the relation between demand and supply of agricultural produce? The FAO has long maintained that to meet the food demand of an increasing population which will become wealthier and more urbanised, agricultural food production will have to increase by 70 per cent from now to 2050. To globally satisfy a diet which will be increasingly enriched in calories and change in its composition, annual cereals production will have to reach around three billion tons, about one-third higher than today, that of soya will have to increase by 140 per cent and that of meat will have to reach 470 million tons, 200 million more than current production.

Concern on the effects of such changes was triggered in concomitance with the sharp rise in farm commodity prices experienced in 2007/2008 and was consolidated as a priority on the international political agenda with the new peak in prices recorded between the end of 2010 and beginning of 2011. Hence the awareness that the equilibrium of agricultural markets risks becoming increasingly precarious due to the structural changes occurring alongside trends in world food demand.

This threat has spawned the debate on food scarcity, on 'how to feed the world in 2050', as runs the title of a recent report by the FAO,[4] whose conclusions are, like those in other studies,[5] very clear and eloquent: winning the war for food security requires urgent action to be taken at the international level. If agricultural production continues to grow at the rate recorded in recent years, the gap between demand and supply runs the risk of widening considerably already in the immediate future, leading to the inevitable extension of the area of food insecurity, which currently exceeds 1.2 billion people, and bringing back onto the agenda, even for advanced economies, the issue of food supply security, which seemed to have been definitively shelved at the end of World War II.

BOX 1.1 THE MILLENNIUM DECLARATION AND THE GOAL OF REDUCING WORLD POVERTY AND HUNGER

The Millennium Declaration was adopted by the United Nations in 2000, with the aim of carrying forward, in coordinated fashion at international level, a series of actions to build a less poor, more sustainable world.

The declaration sets out eight targets for reaching this ambitious proposition, each of which constitutes a 'millennium goal'. The first on the list is that of the fight against extreme poverty and hunger in the world. The undertaking is to at least halve, between 1990 and 2015, the percentage of the world's population who live on less than a dollar a day. The other goals concern the wider diffusion of primary education, gender equality, the reduction in infant mortality, improvement in maternal health, the fight against AIDS and malaria, the Earth's environmental sustainability and promotion of a global partnership for development.

The peaks in food prices recorded in recent years have virtually wiped out the progress achieved on Goal 1 up to 2005. The last report of the United Nations on the state of the Millennium Goals, published in 2011, shows that today the achievement of part of the target results is in doubt or definitively compromised, especially in some areas of the world.

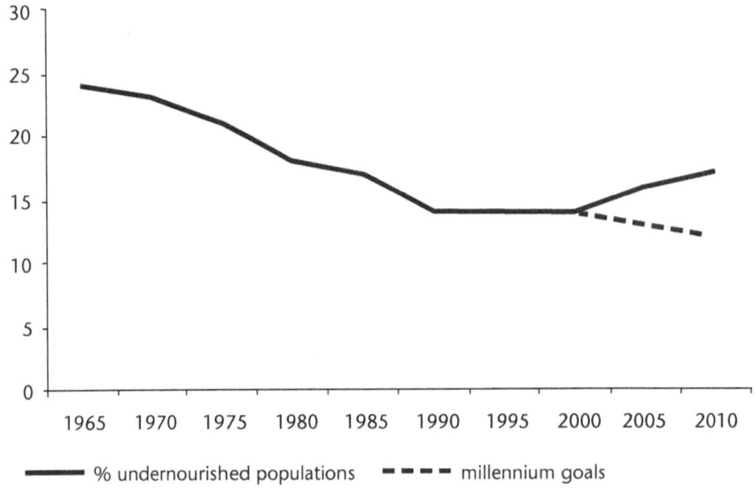

FIGURE 1.4 Heading towards failure in the goal to fight hunger in the world
Source: FAO Hunger Statistics

Stomach and reservoir

To complicate the puzzle of the new world agricultural disorder, biofuels have arrived. Known only by area specialists until a few years ago, fuels to power vehicles obtained from biomass have earned great notoriety in a short period of time. At the beginning they were presented as the panacea for the great problem of climate change, thanks to their capacity to reduce overall greenhouse gas emissions in the atmosphere, replacing petrol and diesel fuel. They later became a 'crime against humanity',[6] insofar as they were held responsible for removing land from food crops and hence for the steep rise in prices in 2007–2008. Finally, they have been recently criticised for their environmental impact, insofar as their rapid development could lead to the destruction of natural ecosystems like forests and peat moors, in turn fundamental as emission sinks.

Thus the matter, which is both complex and topical, has become the subject of a debate that often transcends scientific rigour to take on ideological connotations. After all, there are many interests at stake and it will be fundamental, for the farming and industrial sectors not only of advanced economies but also and especially of poor or emerging countries, that development of this opportunity occurs through maximising benefits and minimising negative impacts.

The use of biofuels in vehicle engines is not at all a novelty from recent decades. At the dawn of the automobile industry, towards the end of the nineteenth century, Nikolaus Otto used ethanol to run his first internal combustion engine, installed on the first car in history, in 1886. A few years later, Rudolph Diesel used peanut oil in the compression which he invented and which still today bears his name. The same Model T Ford, the first car produced on an assembly line, was able to function – what most cars nowadays can't do – with a mix of ethanol and petrol in variable proportions.

Then came the era of cheap oil available in huge quantities and biomass fuels were sidelined, given that there was fuel in nature which was unrivalled in energy power, transportability, ample availability and storage capacity. Nothing changed until the 1970s when, after the first oil supply crisis in history, the world abruptly realised that oil might not always be available or could be very expensive.

It was thus that the first alternative fuel production programmes began to develop, namely bioethanol from sugar cane in Brazil and bioethanol from maize in the US, using agricultural raw materials for which the two countries concerned were and still are leaders in world production. Therefore the element underlying the two more successful biofuel production chains wasn't care for the environment, but a more prosaic requirement of energy provision.

The history of the development of fuels from biomass in the main producer countries is eloquent in this regard. Brazil began to offer state incentives to biofuel producers because their activity was strictly correlated to the sugar sector, one of the most important in the country. Bioethanol is nothing else but the fermentation of sugar cane in alcohol. As Brazil has always been a sugar exporter, it is more greatly affected by possible collapses of international demand. Precisely after a crisis of this type, in concomitance with the depression of the 1930s, the government introduced an obligation to mix 5 per cent with petrol. In other words, surpluses were transformed and reused in civil transport. In the following decades, the international trade scenario improved and such measures were not strengthened. As a result, the chief role of the bioethanol industry long remained that of a sort of small shock absorber for the sugar sector.

However, in the 1970s two things occurred that gave a decisive impetus to the re-birth of the production chain: the 1973 oil crisis was indeed accompanied by a new collapse in international sugar prices. Thus Brazil launched the *Proalcool* programme in 1975, with which bioethanol prices were massively subsidised and the fiscal pressure on fossil fuels was raised, besides creating specific agreements with the main manufacturers to produce vehicles able to run exclusively on ethanol. Production then reached a peak of 9.7 billion litres in 1987, only to enter a period of crisis in the 1990s, when the oil price returned to very low levels and most of the incentives were removed. Yet the production chain was kept alive through the introduction, in 1993, of the 22 per cent mixture obligation of all petrol sold. This policy is still in force.

Also in the case of the US, it was the 1970s oil crisis which triggered a policy of incentivising bioethanol from maize. The first measure was introduced in 1978 through a reduction in the tax for mixes containing at least 10 per cent bioethanol. This was followed by a tax credit for producers. But it is only in the last decade that there has been a decided change in pace, starting from 2002, with the introduction of measures to incentivise biofuel production. Finally, the 2005 Energy Policy Act and the 2007 Energy Independence and Security Act ushered in ambitious objectives in terms of mixture obligations, both with the increasingly widespread bioethanol and nascent biodiesel.

In the rest of the world biofuels have been a talking point only since the 1990s. In Europe, in particular, in 2003 a non-binding mixing objective was established for EU Member States of 5.75 per cent of total fossil fuels, besides the possibility of defiscalising biofuels. With the reform of the Common Agricultural Policy in 2003 production support was also introduced for energy crops. In 2009 with the Directive for the EU's energy strategy the cross bar was raised: the mixing objective was set at 10 per cent for 2020 and is binding. These measures have led to the development of a production chain for biodiesel from rapeseed oil and, to a lesser extent, bioethanol from wheat, maize and sugar beet.

In China and India, biofuels have for some years been the subject of incentive policies. In China, for example, in ten provinces there is a mixing obligation of 10 per cent of bioethanol. On the other side of the Himalayas, India has set a national goal of 20 per cent mixing with biofuels for 2017.

Thus policy incentives have literally created biofuel production chains. Even in a period of very high oil prices, the competitiveness of biofuels without incentives is not yet reality, except perhaps for Brazilian bioethanol.

Currently, the world production scenario is fairly concentrated. As regards bioethanol, the US, with its 42.9 billion litres produced on average in the three-year period 2008–2010, has overtaken Brazil, with its annual production of 26.1 billion litres. Well behind are the other players, with China standing at 7.2 (of which only 2 billion are used as fuel), EU-27 with 5.7 billion litres, India with 1.9 and Canada with 1.5. In biodiesel the leader is the EU-27, with 9.2 billion litres, followed by the US with 1.7, and by Argentina and Brazil with 1.6. Many of these countries and probably others that could follow their example have indicated some key growth targets in biofuels from now to 2020.

The forecasts indicate a growth in bioethanol production for the next ten years of about 60 billion litres, with the US, Brazil and EU-27 soaking up much of the expected increase.

Also as regards biodiesel, an increase is expected of over 25 billion litres. The raw materials used will largely be *ad hoc* crops. In 2020, 13 per cent of global maize production, 15 per cent of vegetable oils and 30 per cent of sugar cane will be used for the production of biofuels. Crops for biofuels have taken up about 20–22 million hectares in the last few years and could, due to further development, reach over 35–40 million hectares in 2020.

When it comes to yields and the consequent impact on the areas used for biofuel production, it is worth stressing that the choice of raw materials clearly conditions the number of overall hectares to be used. For bioethanol, production ranges from 6,000 litres per hectare of sugar cane to 3,500 from a hectare of maize, while as regards biodiesel, production ranges from 5,500 per hectare of palm to 1,200 of rapeseed oil. Producing all the bioethanol by using only sugar cane would require half the number of hectares that would be needed to produce bioethanol only from cereals. The same argument is even more marked if we compare palm oil with oleaginous crops in the case of biodiesel. It is no accident that the EU, in which biofuels can chiefly be produced at low yields per hectare (bioethanol from cereals and biodiesel from rapeseed oil), has explicitly established from now on that to reach its mixing objectives it will have to import considerable quantities of bioethanol, presumably from Brazil, and biodiesel, presumably from Malaysia or from other countries in south-east Asia. With the yields that can be achieved in their own land area, it would take about 18 million hectares, which amounts to about 15 per cent of agricultural land in Europe.

Faced with a rapidly changing scenario, the question that the world has been asking for some time – at least since the 2007–2008 crisis in farm commodity prices – is whether this development is sustainable and what contribution it has made to the price shocks. The policies supporting biofuel production are affecting the food market in two main ways: the change in land use from food to energy production, and the quantity of

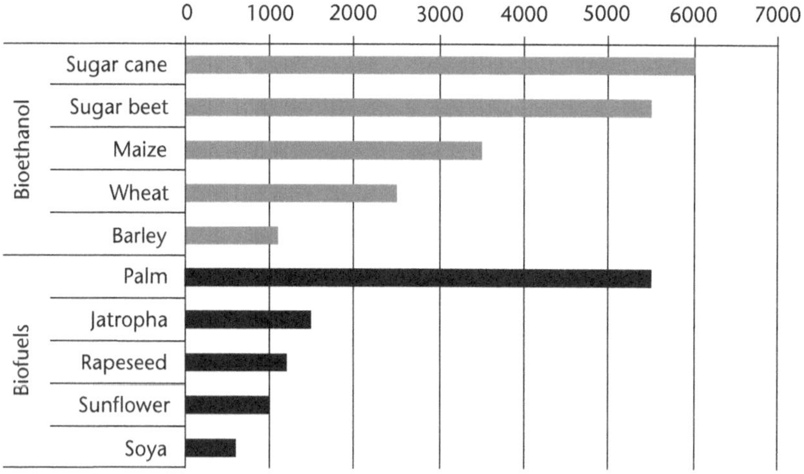

FIGURE 1.5 Yield (litres/hectare) of the main crops used for biofuel production
Source: FAO and WI – Worldwatch Institute

foodstuffs progressively removed from market trading. To what extent this has actually supported the upward trend in prices remains uncertain, yet it is undeniable that biofuel represents an additional source of demand for a market already short of strategic commodities like sugar and cereals.

That said, should the financing and research on the uses of biomass as energy sources be abandoned? Obviously not. Research into so-called second generation biofuels could soon permit the use of cellulose biomass as a production base, the most abundant material on Earth: crop residues like straw and cereal stubble, woody materials obtainable from woodland maintenance, discard from industrial processes like beet pulp, the shells of dried fruit, the organic components of solid urban waste are all raw materials which could soon be used without using a single farm hectare. Needless to say, in terms of environmental benefits, second generation biofuels promise to cut greenhouse gas emissions with respect to fossil fuels by between 70 and 90 per cent.

That of biofuels is the first example which shows us that the path of innovation is indispensable to identify solutions able to ensure greater environmental, economic and social sustainability.

One revolution to complete and one to build

Farm production required for food production may be increased only by following two paths: an increase in agricultural areas and a growth in yield. In the last few years, production increases are to be attributed almost entirely to the latter. Between 1998 and 2008, on an agricultural area which remained almost constant (+6 per cent) worldwide, productivity increased by over 110 per cent.

Besides, the prospect of expanding agricultural areas can today play a marginal role in the objective of increasing the amount of foodstuffs available on the markets. We showed that historic competition between town and country, just like the more recent competition between food crops and energy crops, is compromising agriculture in the world's more fertile areas. Of the land remaining, the more fertile areas are already farmed and thus possible expansion can only rely on marginal and scarcely productive lands or on reduction, not at all desirable, of the current surface area given over to forests.

The per capita farmed area decreased by over 50 per cent between 1963 and 2008. The amount of farmland itself is fairly stable. The used agricultural area (UAA) increased at an average rate of 0.30 per cent per year in the last 40 years while increases close to zero are forecast for the near future (+0.1 per cent per year). This timid expansion is affecting especially the world's more needy areas, which are regrettably also the least productive. Here the increase in the UAA will be around 16 per cent in coming years, against a continued, yet accelerating, decline in available land in advanced and emerging economies (from -0.15 per cent in the period 1967–2005 to -0.23 per cent from 2005–2050).

If the increase in farmland cannot be the solution to the problem, the only way to increase the supply of foodstuffs is the route of technical efficiency. Let us try to understand how we are tackling this point.

We have experienced a period of great agricultural capacity almost on a global scale, called 'the Green Revolution'. This is

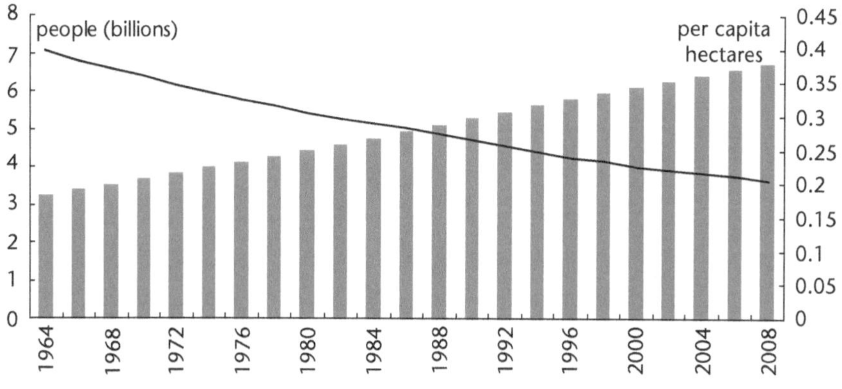

FIGURE 1.6 Trend in population and per capita land area 1963–2008
Source: FAO

a term used to refer to the remarkable increase in world agricultural production between the early 1960s and the end of the 1980s. The technological progress experienced in the more developed economies and its transfer to other regions, especially Asia and Latin America, led at that moment to the doubling of yields for some cereal crops which are basic for the human diet, like rice, wheat and maize, besides the productivity of other plant species and livestock.

This real production boom has been able to meet rising food demand in the past 30 years and has also ensured a certain stability to food prices. A historical phase in which giant steps taken in developing pesticides and fertilisers, combined with an improvement in production techniques and plant breeding, allowed a rapid growth in farmland yields. Certainly, this was not the case everywhere: in Asia, yield increases were obtained practically without increases in cultivated areas, while in Africa yields were stationary despite the bringing into cultivation of new lands.

However, we are talking of a quarter-century of sustained yield increases, coinciding with a marked rise in public investment in agricultural research, both in more developed and in developing countries. The result was a mean annual global increase in cereal production of 2 per cent, with the highest rises in Asia (+2.5 per cent). The Green Revolution seemed universally to be the death knell of the Malthusian theory of growth.

What has changed? How can we account for an alarming situation that today goes well beyond the ever neglected issue of hunger in the world's poor areas? What has compromised the reaching of internationally agreed goals regarding the war against malnutrition? What is lending new vigour to the theories of the Anglican clergyman which we referred to at the start of our reflections?

We are faced with two types of limits: on the one hand the need for more sustainable agriculture, hence based on less use of chemical inputs, one of the main protagonists of the Green Revolution; on the other, the concern at having reached a technological barrier such as to be able to achieve only marginal short-term increases, thereby losing the contribution of what was the driving force behind the productivity increase.

With regard to the former issue, the close link between farming and the environment appears evident: a relation in which agriculture gives positive returns (area and hydrogeological protection, stewardship, sequestering of carbon dioxide), but also negative returns (water pollution, greenhouse gas emission). The latter involves a short-term challenge to sustainability due to a constant process of intensification, which has chiefly affected emerging countries.

As regards the aspect of farm productivity, concerns arise from objectively alarming data, which mark a sharp slowdown in yields in recent years. This is a trend which has dramatically occurred in several areas like Europe, where the mean annual growth rate has fallen from 2.1 per cent to below 1 per cent; in others, as in the case of Asian countries, the slowdown was gentler, though yet of non-negligible proportions.

The OECD and FAO recently estimated that for the coming years the annual growth in production will continue to be slower than in the past, falling from an annual average of 2.4 per cent for the previous decade to 1.7 per cent for the coming years.

According to many analysts, these data indicate unequivocally the end of the season of the Green Revolution and the reaching of a level of efficiency that will be hard to beat in the short–medium term. However, the matter is both very hotly debated and complex. We can say that all agree on one fact: technological progress and its transfer continue to represent one of the main instruments, if not *the* main instrument, to meet the challenge of food security.

There are those who think that in this field little may be expected in the immediate future and others who maintain that research, development and transfer of the innovation may represent not only in the long term, but also in the short term, decisive factors to improve production capacity, without threatening environmental sustainability.

We who are trying to be on the side of the optimists maintain this second perspective. First, because we believe that the Green Revolution is anything but over. Some recently produced data from a study commissioned by the UK Government, entitled *The Future of Food and Farming: Challenges and Choices for Global Sustainability*,[7] show that, in reality, there is still important room for manoeuvre that may be exploited to further extend the benefits of technical progress made until now. There are still too great differences in agricultural yields between geographical areas that are not dissimilar in terms of physical and climatic conditions: particularly marked differences for the world's poorer countries, which are penalised in their access to innovation also by the lack of infrastructures and services. With the removal of these barriers, important new spaces for the growth of agriculture may be created, facilitating the adaptation of technical progress to various geographical conditions. Indeed, innovation needs to adapt to be functional in the economic and structural context in which it is inserted. Innovations constructed for wide scales of production need to have a sustainable opportunity cost even where production structures are small, as often happens in poor economies. This requires an organisational improvement, integration of the agricultural fabric, development of know-how and skills.

According to the study in question, if more advanced knowledge in the technical field were to find widespread application, in many areas in Africa, but also in Asia, production could more than double. This is clearly a situation one would hope for, though it is unlikely to be achieved in the medium term: a prospect that has to factor in historically unresolved problems, like the poor effectiveness of the current system of investments and aid for the development of poor countries and the eternal situation of conflict and political instability of some large areas. This is a prospect which nevertheless suggests that some productivity gains may be achieved in extensive areas of the world, with very important results in terms of the war on hunger.

Achievement of this prospect would, however, be but part of the solution to the problem. To satisfy growing demand requirements, more would have to be produced also in the areas that are today at peak production efficiency and in which, as we have seen, yield increases are dropping off.

What is also in net decline is public expenditure devoted to research and development in the farm sector, which has stagnated in poorer countries, while in more developed countries it is growing at decidedly lower rates than in previous decades. Only in some emerging countries, led by China, Brazil and India, are investments in research and development, albeit decreasing with less intensity than in the past, still very important. The Brazilian agency for research in agriculture, EMBRAPA, is today the largest in the world with an annual budget over one billion USD. In the last ten years China has increased its expenditure on research in agriculture by 10 per cent per annum, falling just short of 4 billion USD in 2008. For the five-year period 2007–2012, New Delhi has allocated over 2.6 billion USD to the sector. Chinese and Indian investments now account for over 70 per cent of total public expenditure in agriculture in the whole continent of Asia.

Many analysts and commentators have pointed out the close relationship between the trend in public expenditure in research and the decline in growth rates for farm productivity. By the same token, the current financial and organisational inadequacy of the system of research in African countries does not allow swift capitalisation of the benefits of technical progress reached in agriculture.

We are dealing with a sector in which, more than in others, public investments are fundamental. The particular fragmentation in much of the agricultural productive fabric limits private research and development activity. Even where firms reach a significant size, such as in the seed or agro-chemicals sectors, they are characterised by a short–medium term view and are calibrated on production scales such as to maximise the rate of return on investments as rapidly as possible. Instead, public research allows investment in long-term solutions, taking due account of the diversity of contexts in which the research is to be applied as well as the social and environmental sustainability of the technical improvement.

The reasons that have discouraged public research in recent years should be sought in the scenario of market stability which we have left just behind us. The slow but constant decline in the prices of foodstuffs, which were a hallmark of a fifty-year period of overproduction, encouraged by policies of sectoral support adopted in the more developed countries, has virtually wiped out the concerns of wealthier nations with their own food security. Now that the age of abundance, so it appears, is over, we need to re-start with public investment in research to tackle the challenges of a new era, that of scarcity, in which to construct a second green revolution.

Regrettably, it should be pointed out that when we talk of innovation in agriculture, the debate often ends up being polarised on two issues of another nature, crystallising ethically irreconcilable positions, as occurs in biotechnology or animal cloning. On the one hand, this lessens the important role played by modern agronomic and agro-ecological sciences in promoting productivity and sustainability, contributing to making them a sort of second-rate science, with less appeal in media terms and as regards investment. On the other, it risks fuelling social fears which in many cases are entirely unjustified.

In Europe for example, there is a *de facto* moratorium on the cultivation of modified organisms with the recombinant DNA technique, known as genetically modified organisms (GMOs). These are techniques that enable work done by the world's farmers over centuries to be carried out with high precision and great efficiency, selecting the most productive and resistant cultivated varieties and creating new ones. Maintaining this moratorium is due to widespread sensitivity that perceives agricultural products as the result of transgenesis which is potentially damaging to health and/or the environment. Thus the progress made by science is ignored: assessments of public agencies that certify the safety of GMOs and the experience gained in scientific research are ignored, or at best classified at dubious.

Certainly, whether GMOs should be used on a vast scale in Europe is open to debate, especially given the need to preserve a certain type of agricultural heritage. Yet the serious thing is that the fear of GMOs is slowing down, and in some cases almost prohibiting *de facto* (as in Italy), research on this topic: precisely in one of those fields that require greater public participation, for a reason that we shall dare to call 'ethical', i.e. avoiding that the poor and smallholders continue to be excluded from the benefits being achieved with private research due to adaptability barriers (hence profitability) of private research to maximise their return through economies of scale. It may be no coincidence that the only GMO which up till now has been extensively used in Africa has been cotton, probably the most organised, efficient and profitable cropping system on the continent, but also the one not used for producing food.

2

NATURE CALLING

The value of diversity

The relationship between agriculture and ecosystems is shown in all its complexity in the sphere of biodiversity conservation. While farming consumes biological diversity, it is also capable of integrating it and creating new diversity, so-called *agro-biodiversity*, the result of synergies between man and nature. The above relationship between farming and ecosystem variability is not linear. It is simply necessary: without biodiversity there would be no agriculture. Its conservation also has direct effects on food production. Biological diversity ensures broad possibilities of crop diversification and greater variability in diet: to cite only one example, the protein content of rice may lie between 5 and 14 per cent, according to the variety that is grown. Conversely, the loss of biological diversity leads to the extinction or marginalisation of plant and animal species, alters the functionality of ecosystems and also has a considerable economic impact. According to one study a few years ago on ecosystem economics, the value of terrestrial biodiversity loss hovers around 50 billion euros a year, 'i.e. a cumulative biodiversity loss of 545 billion euros from 2000 to 2010, little less than 1 per cent of global GDP'.[1]

According to the FAO, since the dawn of agriculture about 7,000 species from 27,000 higher plants have been grown, although today a mere 30 crops supply 90 per cent of the world population's food energy needs. Wheat, rice and maize alone account for almost half. Of the 15,000 species of birds and mammals, only about 30 are used to produce food and less than 14 account for 90 per cent of world livestock production. A characteristic common both to plants and to animals used for food production has been their ability to adapt which, thanks to selection made by man, has allowed them to prosper in a broad spectrum of environmental conditions. However, in recent years, experts have reported erosion of genetic diversity within these species – of that very resource which is especially important for agriculture, since its availability increases the possibility of selecting traits such as resistance to frost or high temperatures, to disease or to parasites.

In general, it is calculated that human activities lead to a rate of biodiversity loss 1,000 times higher than the natural rate. As regards agriculture, the use of intensive monocultures and the massive use of fertilisers and pesticides leads to a simplification of ecosystems which, in the long term, compromises the very capacity to produce. The problem is that traditionally the farming methods with more respect for biodiversity are also those which are less productive, hence less able to respond to the new challenges of food supply. Between 1961 and 1999 the expansion of farmed soils contributed only 29 per cent to the increase in production, while yield increases on land already cultivated amounted to 71 per cent. Moreover, to increase volumes, traditional environmentally respectful methods require larger tracts of land, an option, as we have seen, which is difficult to carry through. On the other hand, intensive methods consume less soil as regards land area, but erode potential soil nutrients more rapidly. This is a double constraint par excellence: we have to produce more for an increasing global population that is changing its diet; at the same time we have to pollute less. Otherwise, it will be impossible to ensure the sustainability of production processes and food supply for future generations.

How do we solve this conundrum? A concept that has gained ground in recent years is that of sustainable intensification in agriculture. As for other sectors of the economy, the chief action to take would be to assume environmental impact as an integral part of production processes in agriculture. The FAO has sought to summarise some of the practices of sustainable intensification, which are none other than components of the broader concept of efficiency, producing with the minimum use of resources. The practices identified by the FAO include a more rational use of nutrients, water and energy, more attention to the conservation of water and soil resources, more use of biological resources in pest control (so-called integrated pest control is coming up with good results on this front), a revaluation of the capital stock of local knowledge with the rediscovery of often abandoned varieties, which may offer solutions for the livelihood of farmers in rural areas in less developed countries and constitute a new stimulus for research and genetic selection, besides having less environmental impact.

The consistency of water

Subjected to growing competition between the various uses, water has become a critical factor – not because its overall availability worldwide is not sufficient to cover demand, but because it is not distributed on the basis of the various regional needs: 15 per cent of the world's fresh water is concentrated in the Amazon forest inhabited by only 1 per cent of the world's population. By contrast in China, which accounts for 20 per cent of the population, only 7 per cent of water is available. This makes the water problem geographically specific. According to the United Nations, by investing 0.16 per cent of global GDP in the water sector and strengthening local supply systems, one could reduce water scarcity and halve the number of those with no access to safe sources in less than four years.

Critical situations concerning water consumption may arise from overexploitation, climatic stress and pollution, in which water courses receive more waste than they can assimilate. In the course of time, some such situations have been alleviated in many parts of the world thanks to technological progress, which has allowed an increase of about 700 per cent in the storage capacity of fluvial systems in the past 50 years, promoting economic development, especially agriculture. By contrast, problems have become more serious in other areas: for example where economic and urban expansion have required large quantities of water, exchanged for ever greater doses of pollution. There are also cases where erosion has led to a reduction in water availability in extensive areas of Africa.

Thus, in many areas worldwide, the equilibrium between the hydrological cycle and the natural system has become more precarious. This trend is interwoven with growth in population, consumption and polluting waste, and is leading to a broadening of the population range affected by water scarcity problems which, according to recent estimates, will rise from the current 700 million people to over 2.5 billion in 2035: that is one out of every three of the world's inhabitants. We leave aside the problem of drinking water, whose scarcity according to the UN already today affects over one billion people.

In the last twenty years, water consumption has increased almost ten-fold, growing twice as rapidly as population. It is now quite clear that we are dealing with a scarce resource. On the one hand, this will make competition for its use even fiercer; on the other, it will be a major constraint for the expansion of all production activity. In particular, agriculture consumes about 70 per cent of the total and will increase its requirement in 2050 by 30 to 50 per cent. Within the perspective of an increase in food demand, it must be borne in mind that to produce a kilo of rice or poultry, about 3,500 litres of water are required.

This is a clear constraint to possible expansion in the agriculture sector, which becomes even more serious if we view the growing negative implications accompanying the relationship between water and agriculture starting with the Green Revolution: agriculture is indirectly responsible for about 40 per cent of pollution of surface water, stemming from the increase in use of chemicals, as fertilisers and pesticides, whose use seems to intensify when areas are devoted to biofuel production.

Of course, there is research that aims to improve crop yields without compromising water and environmental quality. Clearly, such research also has to be incentivised by suitable policies, which are necessary every time that human activity, especially economic, produces negative effects on society. A good example is European law which has permitted the construction of perhaps the world's most sustainable system in the use of chemicals in agriculture:[2] a legal framework designed to protect the environment and consumers, that sets limits and establishes rigorous controls to the use of chemical inputs in agriculture, has promoted the development of agronomic techniques and products for agriculture that have enabled yields to continue to increase whilst abating impacts on the environment and especially on water quality.

Along the path of technical progress an important cog in the wheel to reduce the water requirement of agriculture is the sharing and transfer of modern irrigation techniques, which can allow huge improvements to be made in water savings in many areas of the world. Many responses to this challenge are also expected from crop improvements. To quote only one example, aiming at this result is the initiative led by the African Agricultural Technology Fund (AATF), funded by the Bill and Melinda Gates and Howard G. Buffett Foundations, which is working to produce drought-resistant maize seeds, which should soon be available for small African farmers.[3]

The climatic 'unknown'

Climate change has to be included as one of the constraints to the increase in agricultural productivity. According to the estimates of the Intergovernmental Panel on Climate Change (IPCC),[4] from now until the year 2100 global warming will lead to an average temperature rise of between 1.4 and 5°C. This means that agriculture in many areas in the world will be more vulnerable to water stress, more exposed to the frequency of calamitous events (drought and flooding in particular) and to the action of parasites and other pathogens.

This is due to the effect of the increase in atmospheric CO_2 concentration which is already affecting the frequency and intensity of extreme weather events, and which in the long term could lead to the expansion of desertification, precisely in the world's less developed areas. Nowadays, desertification already affects 100 countries, with about two billion people living in arid and semi-arid regions in the south of the world.

Agriculture is not only influenced by, but in turn influences, climate, insofar as it is responsible for greenhouse gas emissions. Since the beginning of industrial development, the concentration of CO_2 has increased with ever greater intensity. Suffice it to think that, according to data supplied by the World Resources Institute,[5] in the last 250 years over 60 per cent of the overall increase in atmospheric CO_2 levels has occurred in the last 30 years, with all economic sectors playing their part, from energy and transport which account for over 60 per cent, to agriculture, currently responsible for about 13 per cent of global emissions. This is an important slice of the total, placing a major onus on the sector in seeking to attain goals in the war on climate change.

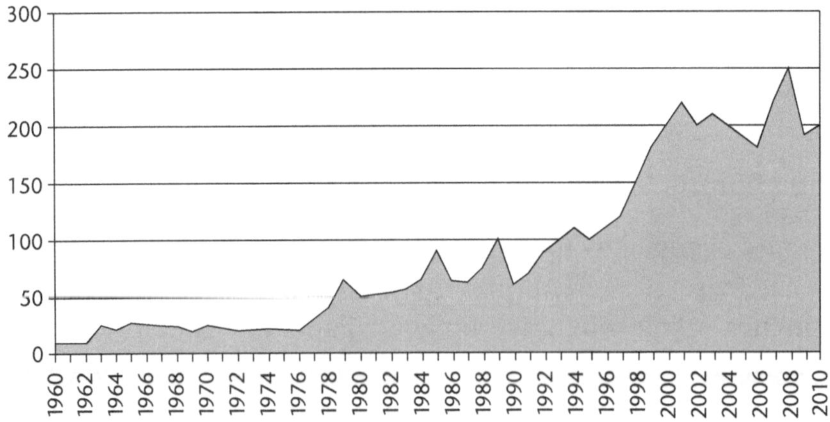

FIGURE 2.1 Number of calamitous events (drought, floods and extreme temperatures) 1960–2008
Source: Elaboration from the International disaster database (www.emdat.be)

Environmental policies which have been developed at the global level from Kyoto onward have introduced ambitious emissions reduction targets also for agriculture.[6] This will squeeze the margins of manoeuvre through which to pursue productivity increases in the near future.

However, in the overall calculation of the impact we should also recall the positive items to be ascribed to agriculture, and more generally, the activity of CO_2 sequestration which takes place in every plant organism. This chiefly applies to pastureland or fruit trees whose ecological footprint is, in most cases, more than positive.

Adaptation to climate change and the contemporaneous promotion of environmental benefits linked to farming thus represent a further challenge for research and policy-making. The efforts of the former will need to be consistent with environmental policy objectives so as to enable agriculture to cope with new, more difficult climatic conditions. The latter will have to be endowed with instruments to incentivise farmers to raise the level of environmental benefits produced, compensating those added costs inevitably linked to the objective.

An unsustainable waste

In an era in which concern for the stability of food supply for millions of people and the environmental sustainability of agricultural production processes become challenges to be tackled with ever greater urgency, there is one fact that cannot be ignored: one third of the food produced worldwide, i.e. 1.3 billion tons per year, is lost or wasted. It is part of the problem of production efficiency, usually given less consideration than saving in input use, energy performance or emissions level, but which has a weight of equal importance. Food waste is tantamount to a futile consumption of resources used to produce it, namely soil, water and energy, and generates unnecessary CO_2 emissions. Each ton of food waste produces 4.2 tons of CO_2.

Inefficiency of the production phases is, however, only part of a question which also affects styles of life and consumption. According to FAO estimates, dispersion (i.e. total loss) of edible matter per citizen in Europe and North America amounts to 280–300 kilos per annum, while dispersion in sub-Saharan Africa and in south-east Asia falls to 120–170 kilos. These figures are not only an expression of higher per capita food production in high-income economies (900 kilos per year) compared with developing countries (460 kilos per year). To better appreciate the phenomenon, we should consider it in terms of the supply chain, the agri-food chain, since food dispersion is distributed very differently according to latitude and individual income. It can be appropriately called *waste* in industrialised countries, where it reaches 40 per cent in the production phases downstream, i.e. in distribution and consumption. The opposite occurs in weaker economies, in which 40 per cent of loss occurs in the upstream phases, that is, in harvest and product processing. Consumers in Europe and North America waste from 59 to 115 kilos of food every year, those in sub-Saharan Africa and south-east Asia from 6 to 11 kilos. The result is that in one year consumers in industrialised countries throw away 222 million tons of food, almost the same quantity of total net food production in sub-Saharan Africa, which amounts to 230 million tons.

In low-income countries, food losses are chiefly due to a lack of efficiency. In this context, a slight improvement in the dispersion level of equipment, energy and waste could have a significant impact on the livelihoods of farmers living on the margins of food insecurity. Dispersion originates in financial constraints, in harvest techniques, in the lack of adequate storage and chilling technologies for difficult climatic conditions, and in shortcomings in infrastructures and distribution systems. One of the possible responses lies in creating 'support, including investment in science and research, technology, education, dissemination and innovation in agriculture, in the distribution and supply chain and among consumers to reduce food waste',[7] but also in improving the efficiency of innovation transfer models, an option which could be supported by aggregation incentives for small farmers.

In high-income economies, waste is chiefly linked to a widespread cultural attitude, a perception of the abundance which legitimates rather irresponsible behaviour in consumption in general, not only in food consumption: a trend that often ends up conditioning contracts stipulated by actors in the production chain. This applies to agreements between retailers and producers which envisage the waste of edible matter only on the basis of product shape or appearance, as regularly occurs in Europe and North America in fruit and vegetables. The 'rejection of waste',[8] the mental elimination of the problem of waste which pervades developed societies, may be overcome by integrating the war on waste into the whole product life cycle, working on awareness on several levels, in the world of production and the retail trade, as well as consumption-linked behaviour. In the past, in the era of surpluses that was the hallmark of European and US food production especially in the 1980s, the EU launched a Programme of Food Aid for the Most Deprived Persons, with the aim of making public expenditure more efficient, earmarking the farm surplus to processing and distribution in favour of the poorer members of the EU. With the end of the era of abundance, the programme now needs an in-depth review, as also established by a sentence of the European Court of Justice.[9] Food waste reduction in developed countries now appears as an initiative that may partly enjoy public support, but that mostly depends on the assumption of responsibilities by the actors in the production chain, where minimal adjustments in the forms of organisation and logistics may allow domestic markets to be reached to which to channel the edible surplus, thereby reducing dispersion and activating virtuous cycles of microeconomics. Indeed, one often forgets that uncertainty in food supply is not a problem that only belongs to low-income countries. Over 50 million people in the US alone are at risk of food security, 43 million in the EU. To meet their needs, organisations have arisen drawing inspiration from the Food Bank model, which recover food unsold by the food sector, and use catering channels to distribute it to the needy through a network of stalls throughout the area.[10] More sophisticated systems in logistic terms, such as the Last Minute market which is gaining ground in Italy, work on a local basis with the aim of allowing a bridge between firms that wish to donate unsold food and associations that can pick it up and distribute it.[11] Even entirely profit-making initiatives may contribute to reducing food waste in developed countries. In the UK a network of businesses is being developed, specialising in the low-cost sale of products that have gone beyond the recommended expiry date, the date beyond which a food product begins to lose some quality characteristics but may be consumed without risk.[12]

3

RECKONING WITH INSTABILITY

Sensitive markets

The international markets for agricultural commodities are the stage on which the new scarcity is shown in all its clarity. The roller-coaster ride of commodity prices fully expresses the tensions between demand and supply which are responsible for price rises.

As already noted, one of the main drivers is the headlong growth in food consumption, associated with population growth, but especially with the higher purchasing power among increasingly broad ranges of the population in emerging countries. As early as the mid 1990s, major increases were being witnessed in the demand for some strategic agricultural commodities, such as wheat, whose prices had long been in constant decline. Demand began to rise at an average rate of 2 per cent per annum, especially thanks to China, India and Brazil, able on their own to account for at least one-third of increases in world requirements at that time. There has been a transition from a period of abundance, albeit characterised by problems of unfair distribution of available resources, to one of structural scarcity manifested in a progressive widening of the scissors between food availability and demand against a trend, at first timid and then more marked, of price rises. For many years the change was not perceived or was, at the very least, underestimated.

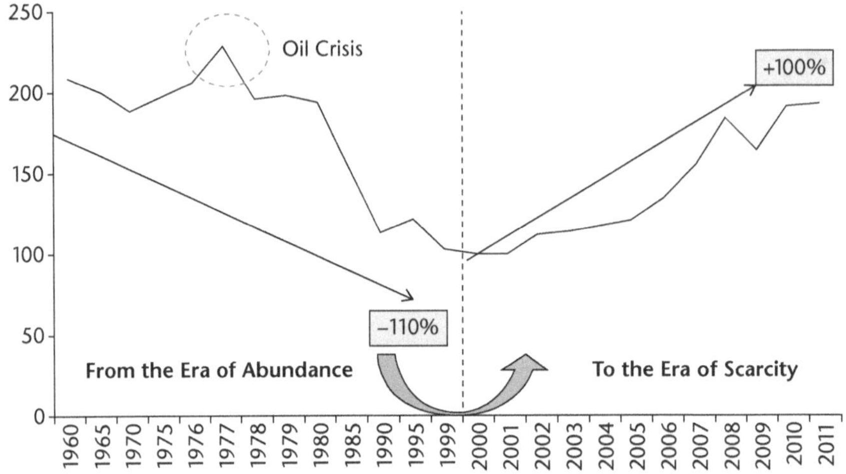

FIGURE 3.1 From abundance to scarcity. Index of real prices of farm commodity 1960–2011
Source: World Bank, 2000=100

Only with the price rises cyclically experienced in recent years has the alarm been raised, especially after the boom of 2007/2008. Although phenomena of this type are not unknown on agricultural markets, the current rises have structurally different causes. The 1970s oil crisis also coincided with a phase of demand increase, but at that time it was chiefly energy prices that pushed up farm prices. Assuming that growth rates of future productivity do not exceed those of the previous decade, the most optimistic scenarios indicate a price rise, in real terms, from now to 2050 of between 75 per cent and 110 per cent for maize, 30–180 per cent for rice and 40–60 per cent for wheat.

What gives cause for concern is not only long-term structural factors. Those of a contingent nature have far more impact than in the past, since the greater integration of the economy favours a more rapid transfer of signals (and shocks) from one market to another. Such integration is also partly responsible for another aspect of the current scenario: market volatility. This term is used when the frequency and range of price variations recorded in a given time span are greater than the historical average. Between 2007 and 2008 the farm price index used by the International Monetary Fund rose by 50 per cent, to then drop sharply (without return to its pre-boom levels) and rise to even higher levels at the beginning of 2011, representing a rise of 130 per cent against 2002 levels.

This phenomenon exacerbates the instability of agricultural markets, which are already structurally turbulent in themselves: commodity prices are often tied to more or less lengthy seasonal cycles – usually annual – separating sowing from harvest. According to the law of demand and supply, the equilibrium price normally tends to rise, and reserves to diminish, in phases that precede the new harvest when quantities become scarce, and to decline, in a good year, in the period immediately after harvest. Thus even small variations in supply or demand can cause significant price movements at certain times.

This natural volatility is exacerbated by a small-scale market trading low volumes and with a small number of exporters, as indicated by the amount of world production of some foodstuffs actually exported. Only 12 per cent of maize and 18 per cent of wheat are traded on international markets; the remaining part stays within producer countries. This means that the quantities traded are very small. Hence, even in the case of modest shocks, repercussions on prices may be significant and the return to equilibrium may take a long time. With such low volumes, concentrated in the hands of a few exporters, small variations in the quantities exported or required for import, produce appreciable increases or decreases in prices.

Clearly, variations in prices are inevitable or, rather, desirable for the proper functioning of markets and efficient allocation of resources. Yet when volatility becomes extreme, it creates a generalised negative impact, especially hitting the poorer sectors of the world's population.

The interaction between climate and energy

According to the experts, one of the manifestations of the phenomenon of climate change is the greater frequency of extreme

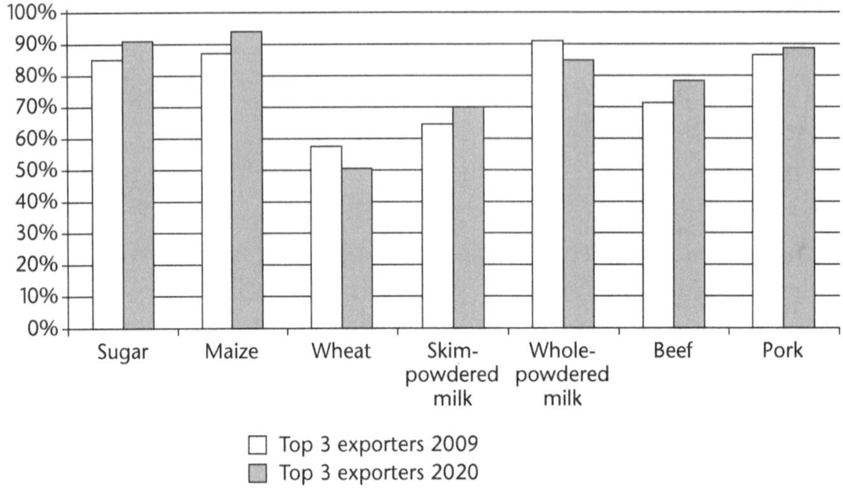

FIGURE 3.2 Market share of the world's first three exporters
Source: Elaboration from FAPRI data

meteorological events. In recent years, these have involved various areas of the world with such intensity that they may be considered among the main causes of farm price increases.

On the receiving end, in 2006–2007, were wheat harvests, which experienced an extraordinary downturn. The drought that hit Australia in those years, one of the world's leading exporters, the heavy rains and the frost experienced by Canada, another leading wheat exporter, actually led to a more than 20 per cent reduction in aggregate supply of wheat from these two countries (which account for over 30 per cent of world exports) compared with 2005 levels. Things were no better in other areas: the drought also hit Russia and Ukraine, and in Europe cereal harvests were almost 10 per cent down between 2005 and 2007, forcing sharp price rises. Of equal impact were the atmospheric events in 2010, from the drought in the US and Russia, where it caused frightening wildfires, to heavy rains in Australia, which further reduced the supply of cereals after the recovery (and consequent drop in prices) in 2009.

Similar dynamics have also affected strategic commodities like maize, soya and, especially, sugar. The poor sugar cane harvests in Brazil and India in 2009 led to a strong reduction in supply and a sudden increase in prices. The good harvests in 2010, accompanied by a slower growth in demand for ethanol due to a phase of economic recession, forced prices downward once again, but only for a few months. They later rose again, supported by the occurrence of adverse climatic conditions in Brazil and Australia, respectively the world's first and third sugar exporter.

While drought, fires, floods and other extreme events have affected crops in certain seasons, they have chiefly amplified the real risk perceived by farmers. Indeed, farming is one of the riskiest occupations due to its close connections with natural events. This condition is made particular by the so-called systemic effect, i.e. the high probability that atmospheric phenomena or negative climatic conditions may hit whole areas more than individual firms, and hence ascribable to the narrow category of risks which cannot be insured by the private market.[1] Especially on small production scales (very common if less than 1 per cent of firms in the world has more than 100 hectares), the repetition of adverse years may lead to the abandonment of agricultural activity and nonetheless contributes to discouraging investment and technical improvement in production units. And this is already occurring in many circumstances.

After all, the very particular risks faced by farmers were the justification for part of the public intervention to support the sector. This has led to the stability of agricultural markets in industrialised countries.

The 'climate' of agricultural markets is made even more uncertain by the strict integration between the market for energy and that for food, which more highly correlated than in the past. This is because the costs of agricultural production are strictly dependent on input such as fertilisers, pesticides, mechanical work and transport, which are obviously correlated to the oil price. According to World Bank estimates, energy costs account for over 25 per cent of the overall cost incurred to produce crops such as maize and soya. This comes as no surprise given that already in the early 1970s the peak of farm commodity prices was chiefly driven by the increase in oil prices.

Between 2002 and 2008 nominal oil prices quadrupled, to then fall after mid 2008, only to record a new boom (+180 per cent) between the beginning of 2009 and mid 2011. This fuelled the volatility of farm prices which react to variations in oil prices according to the intensity of energy inputs used in the production process. A simulation performed by the OECD and FAO shows that increases of 10 per cent in crude oil prices correspond, also given the consequent impetus from the demand for biofuel, to average positive variations in maize prices of around 3.5 per cent.[2]

The role of reserves

The scenario is further complicated by the state of reserves of strategic agricultural products, another aspect that may effectively testify to the transition from a scenario of abundance to one of scarcity. The stockpiling of commodities has indeed played, for a long historical phase, an important role as a buffer for market turbulence as short-term mitigation for seasonal discrepancies between demand and supply.

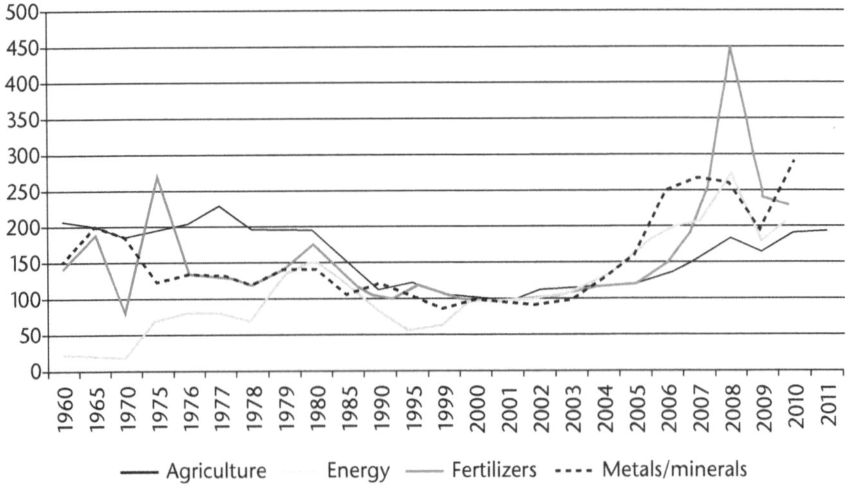

FIGURE 3.3 Correlation between prices of agricultural products, energy and fertilisers
Source: World Bank

Today, the level of food reserves is much lower than in the past. In 2007 cereal reserves reached their historical minimum. This actually made the agricultural supply even more inelastic than it is naturally, further restricting the volumes traded on markets and weakening the capacity to respond to price increases.

In the era of abundance, the long period of falling prices had rendered excessive the opportunity cost of holding large agricultural reserves, traditionally promoted and supported precisely so as to ensure interventions able to re-stabilise markets and protect food security especially in debtor countries. The possibility of finding raw materials at low cost meant that, first in industrialised and then in emerging countries, there was no longer any need to continue to support the customary calming function performed by domestic reserves of agricultural commodities. The same motivations also drove private operators to bring their operative stocks to the functional minimum.

The repeated price booms in recent years have forced governments and private entities to use available reserves to the maximum extent. There is a sensation that it will take some time before cereal stocks can return to suitable levels, especially if natural events continue in the next few years to lead to see-sawing harvests. In 2007/2008, when agricultural production and especially cereal farming were damaged by the adverse weather conditions, the reserves were not sufficient to restore the equilibrium on the markets. The recession-linked decrease in consumption and the recovery in production allowed reserves to be built up once again in 2009, using them to balance, at least in part, the crop losses in the following year. OECD and FAO projections indicate that the greater slowness with which production is rising compared with the past will put a brake, at least in the short–medium term, on the replenishment of reserves, especially if viewed in the framework of the current increase in demand for agricultural foodstuffs.[3]

In addition, we need to rethink the policies designed to fuel food reserves. The function of market management entrusted to the public sector has been considerably pruned down, especially in more developed contexts. This has occurred both to encourage more transparent and efficient functioning of the markets themselves, and due to the misguided conviction of having a future with long-term stability.

One ban too many

The decisions of countries as regards agricultural policy represent a major factor in accounting for tensions on markets in recent years. Government reactions to price booms have aimed to stabilise domestic supply as rapidly as possible by adopting protective measures (such as bans on exports or incentives for imports), to alleviate the impact of increases upon its citizens. Yet these initiatives have had the sole result of exporting instability, taking it from national to international markets, amplifying price oscillations internationally and actually triggering a vicious circle which made the markets even more precarious. This is very widespread behaviour, adopted repeatedly in this phase of uncertainty.

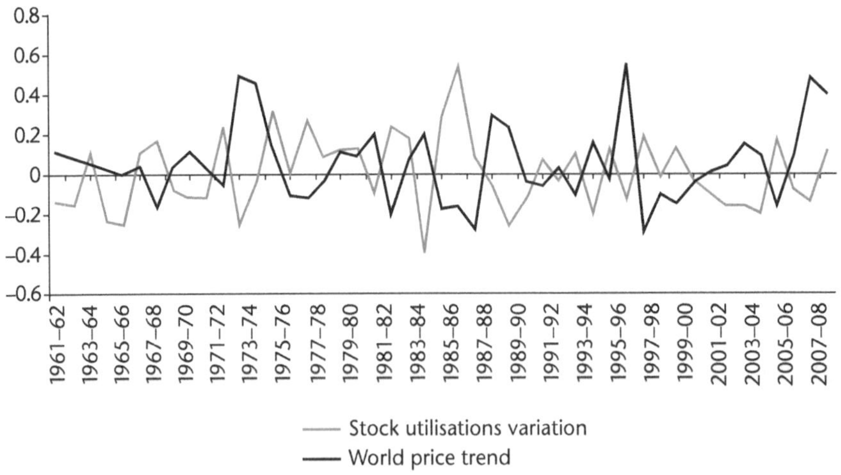

FIGURE 3.4 Negative correlation between maize prices and reserves 1961–2010
Source: World Bank

In particular, it is the measures which limit a country's export capacity which further squeeze market size, driving prices upward: initiatives like export bans or restrictions have been resorted to in the past to preserve the security of national food supplies, but as highlighted by the International Centre for Trade and Sustainable Development (ICTSD),[4] in the last few years have been adopted with much greater frequency, thereby decidedly conditioning the trend in prices on international markets.

In 2007/2008, many exporters of cereals banned or restricted their exports. Ukraine set maximum quantities that were then transformed into outright bans. In Argentina taxation on exports was raised, as in Russia and China, where it reached such levels as to constitute an actual ban. In the same period, major rice producers like India, Indonesia and Vietnam closed down flows to the outside world.

These measures, just like the announcements that preceded them, have obviously played an important role in generating uncertainty: the panic of governments has thus also driven markets into panic. Yet the experience of those two years has not served as a lesson. In 2010 we experienced a new cycle of measures which were completely uncoordinated at the international level. In summer 2010, following huge wildfires which hit Russia and caused damage to crops, Moscow banned grain exports, triggering price increases. The Russian example was followed by Ukraine, while in parallel many governments began to subsidise imports or reduce their taxation. The announcement of the ban was enough for many concerned importers to begin to negotiate higher volumes than in the past, fearing subsequent price rises. According to the FAO, world wheat prices increased by between 60 and 80 per cent between July and September 2010 after the export ban decided by Moscow. Given that these dynamics are no longer the exception but have become consolidated practice, many exporters which previously traded sufficient quantities of agricultural commodities to satisfy a few months' demand today have begun to intervene more massively on markets to cover longer periods.

The effects of this vicious circle also hit the very producers of exporting countries, who see depressed prices on the domestic market whenever they rise on international markets while the cost of inputs continues to rise. This squeezes economic margins and checks growth in production, as testified by the reduction in areas used for wheat farming in Russia, which dropped by about 2.5 per cent in 2011 precisely when international prices continued to rise.

Local solutions without a global vision – this is basically the failing of policies in tackling the volatility of international markets. In this phase the initiatives of *policy-makers* appear completely devoid of coordination, and the very credibility of some international institutions like the World Trade Organization (WTO) risks being questioned. Indeed, WTO rules were conceived to facilitate and liberalise trade in the era of abundance, and attention has focused especially on the need to proscribe attitudes that could impede imports, rather than exports, justifying, with what might be considered superficial with hindsight, reductions or interruptions in export flows by virtue of interests linked to national food security. Until a few years ago, this scenario was considered remote. Only in June 2011 was a proposal presented to the WTO by Egypt to ban restrictions to exports of agricultural products towards poor countries which are net importers of foodstuffs. Subsequently, the G20 group urged the WTO to adopt a specific resolution on export restrictions and bans.

Speculating on speculation?

The role played by the financial markets during the price rises is more hotly debated. Some governments, especially France during its G20 presidency, but also several analysts and representatives of international institutions, have pointed the finger at financial speculation, it being identified as one of the main drivers behind recent booms in farm prices. The accusation of excessive financialisation of farm commodity markets, in the wake of the brouhaha aroused by the financial bubble that preceded the current phase of world economic recession, has had a great impact on public opinion: the activities of investors in the field of agricultural financial markets are increasingly widely labelled as unethical.

Before going into further detail, let us try to explain in a few words what these markets are and how they work.[5] To start with, we are in the field of so-called *derivative finance*, i.e. in markets which trade securities whose value is linked to the price of a certain real good, termed *underlying*. In our case, the underlying security is an agricultural commodity, like wheat or coffee, and its value increases with the rise in the price of such products on the real market, referred to in jargon as the *spot market*. Among the various instruments of derivative finance, perhaps the most commonly cited are *futures*: a contract is stipulated under which two parties (a seller and a purchaser) agree the price and delivery date of a commodity, according to pre-established standards. This means that delivery is postponed, and that at the moment agreed for the transaction, the market price may be different from that agreed upon, and one of the two parties will have earnings from this difference. If the price increases, it is the purchaser who makes a profit. The futures market features two large categories of participants: those who wish to cover themselves from the risk of price collapse, generally farmers, known as *hedgers*, and speculators. The former protect themselves from possible falls in prices by fixing them in advance or purchasing contracts whose quotations follow the presumably opposite trend to those of the good actually possessed. Especially in periods of marked volatility, derivatives are a fundamental risk management instrument for farmers.

By contrast, speculators bet on possible rises. In reality, on expiry of the contract, delivery of the goods rarely occurs, replaced by monetary compensation for the difference between the market price of the commodity on expiry of the contract (*spot price*) and that established by the parties at the time of stipulation. Even if in practice it often becomes very complex to make a clear distinction between hedgers and speculators, the activities set up by the former are generally termed *commercial*, while those of the latter *non-commercial*.[6] Both are indispensable to make the derivatives market work: without the contribution of speculators, there would not be sufficient liquidity.

This mechanism caught the public eye when it was revealed that, in concomitance with price rises, the volumes of derivatives traded on agricultural commodities markets had grown exponentially.[7] Both in the two-year period 2007–2008 and in 2010–2011 there was sharp growth in non-commercial positions on the futures market, led by the activity of investment funds aiming to diversify their portfolios and gain profit from trends in agricultural prices. The key question is whether or not this correlation between prices and investments is 'piloted' by speculative activity.

In its broader, more authoritative strands, the economics literature is rather sceptical regarding the nexus of causality which is thought by many to connect speculation and price rises. The mechanism of financial markets actually seems to be fairly efficient at ensuring that the prices created on commodities markets are consistent with market equilibria. The trend in futures quotations is theoretically tied to expectations on demand–supply relations and thus tends to converge on the real market value of the traded commodity as the contract expiry date approaches.

The stock market should reflect the market and promote transparency of its signals. Moreover, this task is made easier by the standardisation of contracts (the same, clearly defined rules for all operators) and by the presence of a compensation mechanism (technically a *clearing house*) which ensures that the parties respect their obligations assumed, thereby avoiding risks of insolvency. In addition, in the wake of recent financial scandals, the regulations governing transactions on the commodities market have been strengthened. More stringent limits were placed on speculative positions, especially in the USA, the market where by far the largest volume of deals is concentrated. And beyond the emotions that inevitably prevail when one speaks of food prices, it seems totally clear that when there are expectations of rising prices, for structural reasons and otherwise, which we have examined up till now, such expectations are metabolised by the operators, leading to an increase in the volumes traded.

By contrast, when in 2009 harvests were abundant and reserves partly replenished, speculative activity collapsed by over 50 per cent. This faithfully reflected the market situation.

We should point out that, even if there is no overwhelming evidence, there is much doubt concerning the interpretation of reciprocal influences between the real market and financial markets. The only certain way in which this occurs is in trade flows, on the part of individual operators, both on the physical market and on the financial market. This may happen when an owner of extensive food commodities intentionally reduces their supply, purchasing in parallel, large volumes of bonds on the derivatives market. Clearly, this may generate large profits and create market distortions, but it is a practice forbidden by commodity market regulations.[8]

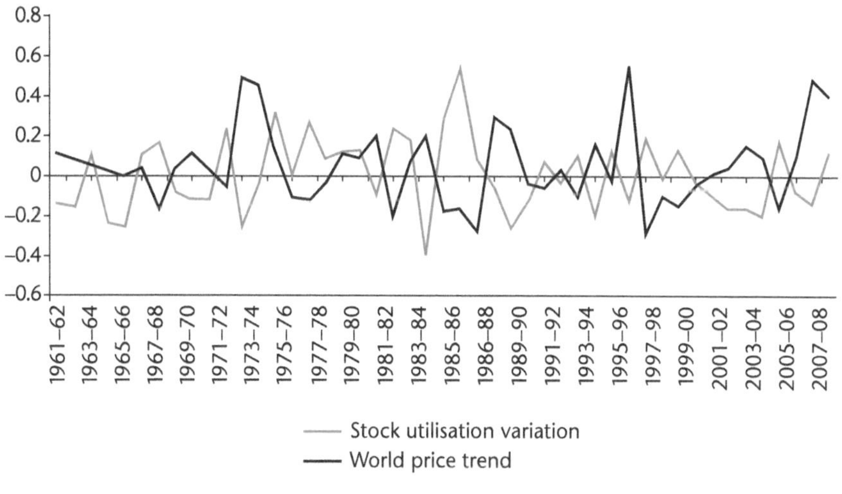

FIGURE 3.5 Number (in millions) of futures contracts on cereals and soya traded on the US market
Source: US Commodity Futures Trading Commission

By contrast, financial operations conducted outside commodity exchange circuits are different, such as in the so-called 'over the counter' (OTC) market,[9] where it is large institutional intermediaries which trade commodities through non-standardised contracts and without solvency guarantees made available by stock exchanges. This generates large risks, like those which became reality during the financial bubble in 2007, in which the contractual reneging of many players who had taken on excessive risks led to real market failure. Today, the volume of traded derivatives in the OTC market in relation to farm commodities is still marginal and we cannot run the risk of it increasing, perhaps as a haven from excessively constraining stock market activity.

Three crises, only one alarm

The early 1970s, as we recalled above, experienced a sudden peak in commodity prices, spurred by factors that only partly coincided with those recently conditioning the trend in farm commodity prices. At that time, a temporary disequilibrium between demand and supply[10] was accompanied by rather low levels of food reserves and especially by a rise in energy prices.[11] Thus, also in 1972–1974 a reduction of less than 2 per cent in world wheat production led to a rise in prices of over 200 per cent. At the time, a major part was played by the sudden rise in oil prices and a precise strategy on the part of some large exporters who, in a phase of falling prices, sought to drive subsidies to farmers upward, thereby reducing the level of reserves and stimulating production diversification programmes which had the effect of reducing the supply of cereals. The area which the world's four main exporters (USA, Canada, Argentina and Australia) used for wheat production fell overall from 52 to 34 million hectares in only four years (1968–1971) and world reserves fell by about one-third. This coincided with a 10 per cent increase in international demand.

However, the recent peaks in farm commodity prices are profoundly different. Most importantly, in the 1970s production growth rates were able to ensure swifter responses to consumption trends. It was right in the middle of the Green Revolution, with its high increases in productivity rates. Today productivity is growing less and demand increasing more rapidly than then. Moreover, the frequency of adverse atmospheric events and higher market integration are playing a more incisive role than in the past.

For example, we may observe that exchange rates, generally expressed in US dollars, have a non-marginal effect on the dynamics of farm commodity prices, as indeed holds for other raw materials. Market signals were extensively distorted both by the appreciation in the euro against the US dollar between 2001 and 2007, and by the exchange rate with the main currencies during the continuing recession. Moreover, today the link between prices of various crops is more robust than in the past. When the price of a product rises, it is swiftly followed by that of possible substitutes (such as cereals and soya in animal feed), even in cases in which production of the latter was abundant in the year in question. Finally, when the price of a *commodity* rises, land is used to increase production, to the detriment of other foodstuffs whose price is destined to rise with the inevitably lower harvest forecasts. By the same token, trends in other commodities (besides energy, metals in particular) also have a greater effect on agricultural markets.

Further, the most recent peak in prices (that of 2010) was partly different from that in 2008. In 2010 a greater role was played by expectations and the consequent reactions of exporters and importers of strategic raw materials. Indeed, in the two years prior to this new boom in prices, production was abundant, chiefly thanks to favourable climatic conditions, and reserves were also at decidedly higher levels than in 2008.

Yet when, in summer 2010, news spread of the great drought in Russia, Ukraine and Kazakhstan, the forecasts on harvests were therefore adjusted downward, and panic gripped the markets. The reactions of governments, both immediate and uncoordinated, triggered a race for cereals which in a short

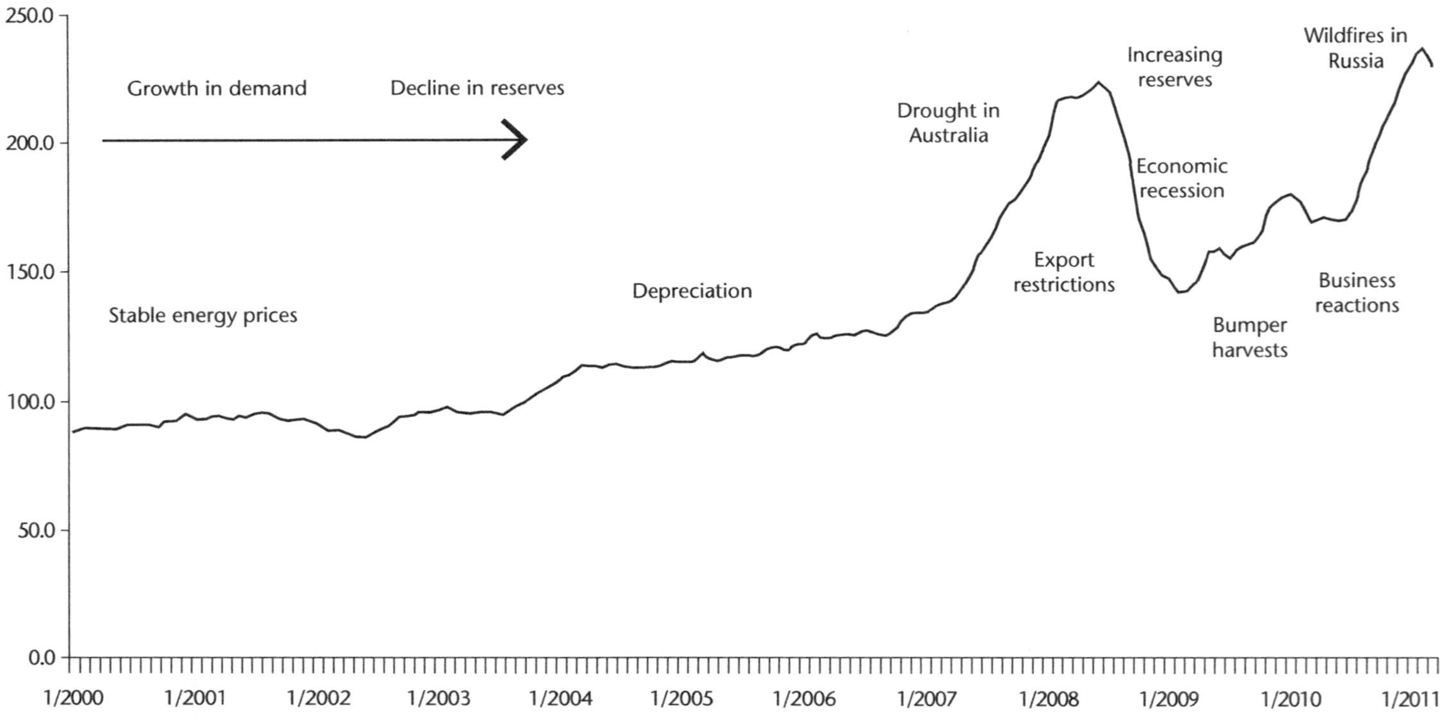

FIGURE 3.6 Synopsis of the main factors that have led to volatility in farm prices in recent years. FAO price index 2002=100
Source: Elaboration from World Bank and USDA data

time, well before the harvests, led prices to shoot upwards, with the contribution of another major devaluation of the US dollar. The exporters closed their borders, the importers sought huge, rapid purchases, and this triggered the sharp rises in commodity prices in the following months, which equalled and exceeded the records reached in 2008.

Who pays for 'agflation'?

The primary sector and agricultural commerce have always been more protected and regulated than other sectors, due to the need to ensure citizens have access to food at reasonable prices. This is the first step on the welfare ladder, the basis for a country's social stability and economic growth.

In recent years this prospect has weakened in many areas worldwide, compromising, as we have seen, the attainment of the primary millennium goals for development. Growth and volatility of farm commodity prices are further compromising the already fragile economic and political situation of many low-income countries. For the most part these are net food importers, which are particularly affected by food price rises. Their impact in such contexts increasingly jeopardises the equilibrium of the domestic balance of trade and leads to unsustainable increases in the cost of living. In the world's more developed areas the share of income allocated for food purchases is around 10 per cent, of which less than 20 per cent is used to pay for the use of farm products. The remaining part goes to cover services that are added (starting from processing) during the product's journey from the field to the table. In less developed areas, the situation is radically different. Families invest much of their income in food purchases, almost all consisting of basic farm products.

If volatility may have a marginal effect on consumption prices of richer countries, for poorer countries the effects of price increases are decidedly more impacting, with consequences such as the increase in poverty and the progressive slipping into the area of malnutrition for millions of individuals and families.

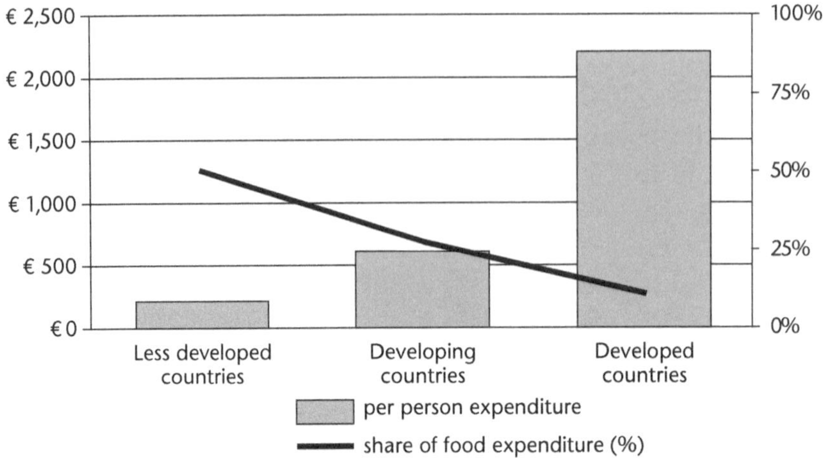

FIGURE 3.7 Food expenditure in absolute value and as a percentage of overall household expenditure
Source: World Bank

Higher supply costs also affect the deterioration of the national balance of trade, on most occasions accompanied by the fall in exchange rates. The result cannot but be a reduction in investments and a slowdown, if not even a fallback, in development.

Yet the effects of the rise in food prices are also felt in less poor countries and emerging economies, where growth is often accompanied by widening of the scissors separating well-being and poverty. In these countries, those who suffer are the still vast group of the less well-off. To indicate the contribution of farm prices to general inflation, a new term has been coined, namely *agflation*, which stems from the combination of the terms *agriculture* and *inflation*.

What is occurring on the markets does not favour, unlike what might be thought, the agricultural producers either, for various reasons.

In general, the structure of the agri-food system penalises farmers, tending to exclude them from the benefits of price rises and offloading onto them the weight of price falls. Price transfer mechanisms are indeed affected by the diversity of structural and organisational conditions affecting the production chain, relegating farmers historically to a so-called position of *pricetakers*, unable, that is, to affect the formation of prices on the market – a condition due to the great fragmentation of the primary sector, while the phases downstream of the production chain, the processing industry and especially the distribution network, are ever more concentrated. This tips the structure of bargaining power in their favour. The same considerations hold for the relationship between agriculture and suppliers of input such as energy and fertilisers, which are essentially multinational companies.

This imbalance does not lead to a rapid or correct transfer of price signals, and penalises farmers. The prices of energy, fertilisers and pesticides rise to a greater extent than those of agricultural commodities, and when the prices of the latter fall, those of input do so more slowly. This ensures that the farmer's margins remain almost unchanged in the case of a rise in prices and that they are significantly squeezed in periods of commodity price declines, according to the use intensity of some inputs. In the case of particular sectors like that of livestock, the complementary effect of a feed cost increase, which often represents the chief component of operative costs, should also be considered.

Price instability, and the climate of uncertainty that arises from it, risk conditioning negatively the trend of investments in the sector. In weaker contexts, but also in ones that are less so, price volatility threatens the very economic sustainability of agriculture. Variations in input and output prices are so poorly synchronised as to frequently lead to a financial deficit, especially in smaller production units with few chances of credit access. This may mean abandonment of farming or downsizing to achieve the sole function of self-consumption for rural families.

Interview with Franz Fischler (Former European Commissioner for Agriculture)

Former EU Commissioner for Agriculture from 1995 to 2004, Franz Fischler is closely monitoring the global dynamics of food supply. A convinced pro-European, his name is linked to the most incisive reform of the Common Agricultural Policy, in the year 2003.

What is happening with the global commodity prices? Is their increase part of a cyclical development or is it a new trend?
We must differentiate between the long-term trend and short-term cyclical or volatile phenomena. In the long-term trend we can observe a change in paradigm. Whereas we saw a permanent downward trend for agricultural commodity prices in the last 40 years, we can expect an upward trend in this decade.

In the OECD, the USDA and also in the EU-Commission-services, they all come to the same conclusion. The drivers of this new trend are primarily growing input costs, e.g. for energy, fertilisers or crop protection, but also an increase in demand. The increase in demand is a result of the growing world population, the interest in more animal protein, mostly in emerging countries, and the use of commodities for fuels and other non-food purposes. The growing price volatility is primarily caused by fluctuating harvest levels as a consequence of climate change, by low stocks and new forms of derivative speculations. So we will see an upward trend overlapped by an enormous price volatility in the future.

What are the effects of this combination of long-term trend and short-term cycles?
The long-term trend clearly favours agricultural production although we should be cautious because most of the output price increase will be absorbed by the higher input prices and therefore the profit will grow little.

The uncertainties caused by growing volatility will negatively affect not only the crop production of farmers but also the other links of the food chains. The biggest harm will be caused for the most vulnerable ones, the small farmers and the poor consumers in developing countries.

Transparency was a key issue addressed by the G20 summit in Paris. What should be done to make agricultural markets more transparent?
The most important instruments to increase the transparency of markets are monitoring and information.

Who should do the monitoring?
Price monitoring is an issue for governments and also for the European Commission. Furthermore, commodity stock exchanges, consumer organisations and retailers can also contribute a lot to more transparent markets.

Are the European commodity stock exchanges strong enough?
Unfortunately not! We in Europe have more than 30 but only small stock exchanges for commodities, and the result is that international price notifications come from New York and Chicago. If we would like to achieve international appreciated price notifications in euro prices then we would need to establish a strong European commodity stock exchange.

The other big issue in Paris was the so called 'financialisaton' of agriculture. How do you see the role of speculation in agriculture?
The traditional forms of speculation, the so called 'Futures Markets', are an important tool to stabilise prices and to help the traders in their business. So, I see this form of speculation as being very positive.

But the new financial derivative speculations strengthen the upward and the downward trend of the price-cycle. I don't think that derivative speculations can create price volatility but they can amplify it. Therefore better rules and disciplines would be useful to limit the harm of these new instruments of the finance industry.

Must these disciplines be introduced at a global stage?
Ideally yes, but I think Europe should not wait. As long as a global agreement is not in sight, Europe should go ahead and pass the necessary legislation.

A few years ago biofuels were considered the salvation from fossil fuels. Nowadays some people say that agrofuels would be a 'crime against humanity'. Should we quit the biofuel option?
The first generation of biofuel production is in the best case a kind of transitional measure. My personal view is that this production is not sustainable and can't therefore be the base for alternative fuels. There can't be the slightest doubt that absolute priority goes to food production and not to fuel production. The only available sources for sustainable non-food use is biomass from forest residues and waste. But even for this use we should respect certain priorities: for example the material use of biomass should be prioritised against fuel production and combustion.

What is material use?
With modern and innovative production techniques we can replace oil and other non-renewable resources in the chemical industry, in house construction and in many other industries. These concepts are part of the 'green industries' and can help to convert our economy to a knowledge-based bio-economy. Therefore Europe should intensify research and development in this field.

Research and Development investments have decreased in recent years. How can we steer innovation and integrate agriculture in a knowledge-based economy?
In my view innovation in agriculture is the key for success in the future. We must establish a kind of innovation chain. This means close cooperation between researchers, technology developers, product designers and entrepreneurs.

Consequently we should also consider new forms of incentives, be they financial or clusters, start ups or other integrated approaches. This requires more interdisciplinarity, more risk capital and also better education and training. We also need more private–public partnerships, pilot projects and partly also a rethinking of our support priorities and the role of public research in the future.

Can you give an example what role the public research sector should play?
We should not forget that private research is mainly driven by profitability and the return of investment. Therefore the private sector concentrates on the most attractive and most promising projects. But there are also other scientific questions of utmost importance. Take the example of crop breeding. In recent years wheat improvement was almost neglected and so the global wheat yields grew much less than the maize yields. The main reason behind this is the resistance of wheat to hybridisation. Here is a duty for public research to fill this gap.

When you speak about breeding you mean GMOs?
No! Crop breeding is a much broader task than just genetic engineering. Especially in the developing countries enormous progress is achievable without GMOs.

In your opinion what are the priorities for achieving global food security and what should Europe contribute to it?
The biggest problem in combating hunger in the world is the vulnerability of the 500 million subsistence farmers. Their problems need a holistic approach based on in-depth analysis of the strengths and weaknesses in a certain region. The vulnerability of peasants can only be reduced by the implementation of strategic regional development plans under the responsibility of the government and in cooperation with international organisations. The industrialised

countries must help to develop better production technologies and must provide the necessary financial means. In the last 20 years the share devoted to agriculture from total development cooperation funding diminished from 20 per cent to 3 per cent. This must be reversed. Another area of better regulation is trade policy. It is a pity that all the promises given in the Doha Development Agenda (the current negotiation round of the World Trade Organisation) are still open. The EU has at least implemented the 'everything but arms'[12] concept and has changed its export practices so that it became a net importer of agricultural goods from developing countries. We must also help these countries with aid for trade to improve their standards and to make more trade happen.

Finally, effective and efficient emergency programmes will still be needed because disaster relief for all forms of natural or economic or political catastrophes should be seen as a human right.

PART II

Everyone's problem

4

FAT COWS, LEAN COWS

The new commercial order

The new scenario involves an alteration in the global balance of the food supply, a change that is proving problematic, not only for the parts of the world historically affected by famine and malnutrition, but also for new players in the global economy, for high-income economies and the food processing industry itself. In this part we are going to try to identify a set of phenomena that highlight the potentially disruptive effects of this scarcity, not trying to create panic, but to show that the issue of food security, rather than belonging to a hypothetical future, is already influencing our present, raising doubts over our capacity for reflection and action and setting us challenges that need urgent and adequate responses.

An analysis of the international trade flows in food helps to better define the immense changes that we are currently going through. In 2009 the international trade in food amounted to 1,169 billion USD, more than double what it was in 1999, when it was 548 billion USD. In just a few years and in spite of a severe economic recession, the trade in farm commodities has accelerated at a pace never seen before in the history of world trade. What is more, the equilibria within this equation have changed substantially. Underlying these changes are differing growth rates in food consumption and the rise in some areas of the world, particularly in Latin America, of new agricultural superpowers such as Brazil and Argentina for whom agricultural commodities make up more than half the total value of the countries' exports. Asia and Africa, on the other hand, are heading towards further expansion of their net agricultural commodity requirements. Even the countries that made up the former Soviet Union are changing their position in the context of international trade and shifting from being net importers to food exporters. It should, however, be remembered that the international trade in foodstuffs currently represents just 8 per cent of total trading volumes and, on average, only 15 per cent of global production is traded on international markets, the largest volumes being oilseeds, of which 57 per cent is traded, sugar, 20 per cent and wheat, 15 per cent. The percentages are very small even in cases of important commodities which represent basic elements in the diet such as rice, for which less than 7 per cent of the total production is traded on world markets.

In addition to the increasingly hefty presence of Latin America in both import and export markets, the ever more important role of Asia, particularly China and India, should not be overlooked, their weight having significantly increased in worldwide trade, especially in the field of imports. These phenomena are certain to grow stronger in the future in the light of the economic prospects of the areas in question. Brazil alone, for example, has increased its net foodstuffs trade surplus balance by 460 per cent over the past ten years and forecasts indicate a further increase of 50 per cent in the coming decade.

In essence, as with other economic sectors, the agri-food trade model, constructed on a North–South axis, in which the key commercial players come from within the more developed world, is giving way: after determining global flows for several years it is gradually being eroded by an increasing share of trading volumes moving South–South, between developing countries, rising from less than 10 per cent of total trading volume during the 1980s to more than double this in 2010. This change in trade balances has been taking place for more than a decade, with a reduction in trade flows from countries with advanced economies to developing ones. This decrease, together with the one that has also affected trade between industrialised countries (Europe and North America), has shifted to the benefit of trade between countries in the developing world along South–South lines and today represents more than a quarter of the value of the world trade in farm commodities.

The weight of growth

The rip-roaring development of emerging economies is not without its contradictions, relating to the sphere of food security. To quote just one figure, the 'agflation' discussed in the previous chapter has grown, on average, by 5 per cent in the advanced economies, but is well into double figures in countries such as Brazil, Russia, Indonesia and China which, in 2010, launched a government plan based primarily on keeping a lid on food costs. With about 20 per cent of the global population China has only 7 per cent of the arable land and an economy that, despite the recession, is showing 9 per cent year-on-year

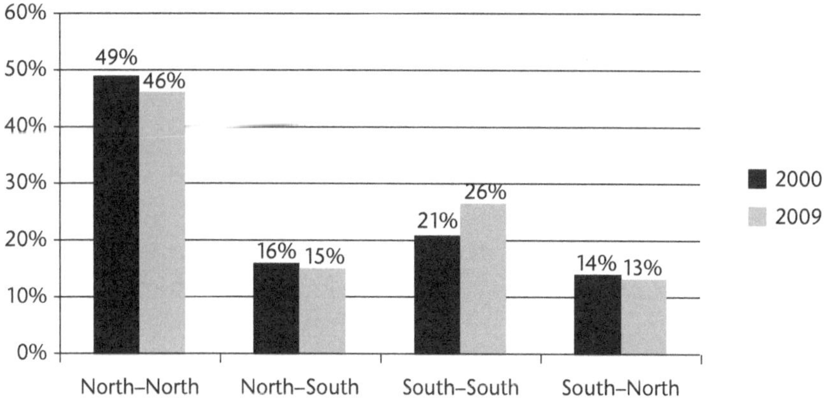

FIGURE 4.1 Changes in the main axes of world trade in agricultural products 2000–2009
Source: Elaboration from FAPRI data

increases in GDP, with food inflation running at around 14 per cent. What's more, the marked inequality between the incomes of those living in urban areas and those living in the countryside has encouraged massive migration from the countryside to the cities over the past thirty years. Urbanisation has eroded the *per capita* availability of agricultural land, which has been reduced to settle at an average figure of 0.1 hectares. These numbers tell of a country hungry for bread and energy, with a rapidly expanding middle class and cities for which food security is a high priority marked by successes and beset by challenges.

With a gigantic effort, between 1991 and 2001 China doubled crop yields, reducing rural poverty by 63 per cent and lifting 400 million people out of poverty by investing in new techniques to increase the productivity of small farmers. Although several areas of the country, especially rural ones, cannot be said to have completely escaped uncertainties in supply, since 2007 China has managed to achieve food self-sufficiency and has proved to be one of the most generous donors to the countries in the Horn of Africa affected by drought in recent years. But the spike in prices in 2007/2008, and even more so that in 2010, raised the bar with regard to the balance between supply and demand. Urbanisation, decisive in the expansion of access to income opportunities for millions of citizens has, as its counterpoint, an abandonment of the countryside of dramatic proportions, which casts dark shadows on the possibilities of the country to support itself. The intense exploitation of the natural resources needed to feed the rapid pace of growth has reached critical levels, particularly in relation to the depletion of water resources. Migration from rural areas to the cities directly impacts on the availability of arable land and on the establishment of new styles of eating. To cope with qualitative changes in diet, which we have already termed a 'substitution effect', last year the Asian giant became a net importer of maize for the first time in 15 years, while the demand for sugar is expected to rise by 50 per cent in 2012 to meet the growing demand for soft drinks. Another critical factor is water. According to the IWMI, the

International Institute for Water Management, the catchment areas in the north, where food production is largely concentrated, are in some cases 'beyond the limits of sustainable agricultural water use'.[1] Since an increase in exploitation by industry and for private consumption is anticipated for the same river basins, there are grave concerns that the required reallocation of resources will actually end up penalising food production in the area. Bringing together all the factors (urbanisation, new eating habits and water scarcity), it seems the lack of water will affect the very regions where harvests will be used to meet the demand for animal protein-based foods. These are precisely those sought by the new middle class, while rice production in the south, where water is plentiful, would ultimately fall. But Beijing, with more than a decade of the struggle against hunger under its belt, is gearing up to meet the new challenges through an extensive network of trade agreements and investment in the production of foodstuffs outside China and in employing its capital in research and development in agriculture within national borders.

Where development does have dramatic implications in the field of food security is in India where, between 1990 and 2005, the main indicators measuring the country's economic growth doubled, whilst, at the same time, 65 million Indians – the same as the population of France – swelled the ranks of the hungry, who, by some estimates, number 200 million people, including 57 million children. All this is taking place in a country where almost half of all children aged five and under are underweight due to lack of food and where most families live on less than two dollars a day. As if inflation on food were not enough, after the last peak in international prices, in 2010, the decrease in people's purchasing power on food reached rates exceeding 15 per cent per year. The balance of payments of New Delhi is very exposed on the market prices for agricultural commodities because following the expansion of consumption, India has begun massive imports of basic foodstuffs such as legumes (+157 per cent between 2004 and 2009) and cooking oil (+68 per cent over the same period).

The issue of food security is not only dependent on market volatility. After a period of heavy investment in new agricultural technologies in the 1960s and 1970s the Indian government has preferred to concentrate its energies on other economic sectors so that agriculture, which employs 52 per cent of the population, ends up contributing less in terms of GDP than one would expect, a mere 15 per cent. On the one hand, the primary sector has stayed at subsistence levels, with millions of farmers who have never seen a modern irrigation system and are unable to guarantee the levels of productivity needed to feed themselves, let alone the Indian people; whilst on the other, the country continues to implement an agro-industrial model with major environmental impacts. New Delhi has repeatedly announced plans to deal with the situation by undertaking a major agrarian reform, which for now, is languishing in Parliament. In the meantime, the government is engaging in investment in agricultural production abroad, mainly through support of projects for large farms with the national governments of developing countries for whom loans and credit facilities are guaranteed. The practice of agricultural outsourcing, including the purchase or long-term lease of cultivable land, through the more or less direct support of national authorities, on the other hand, is being widely used by emerging economies or states with plenty of liquidity but poor in terms of areas which can be used for farming. The consequences of these policies will be examined later in this chapter.

Bread and democracy

'Democracies of bread' is the definition social and political scientists were already giving to the Arab regimes of the Middle and Near East and North Africa in Nasser's day to highlight the subsidies in place for the purchase of bread, the seal of the social contract between rulers and their subjects.[2] It is certainly no coincidence that the 'Arab Spring' started from riots for bread, as a symbol as well as a staple food, triggering a range of social unrest that is shaking the region, with still uncertain outcomes. This is not to reduce the multiple causes of the 'Arab awakening', as it has been termed, to a need to eat. The food crisis, which also hit certain spots in the Arab world hard in 2008 and 2010, cannot satisfactorily explain the complexity of a historical event involving countries that are very different from one another, from Algeria to Syria, not forgetting Libya; differences in context that are expressed in the spark for and motivations behind the protest, in the different interest groups around which the opposition forces coalesce, in the manner in which they work together as well as the action and reaction of the various parties to the events.

By and large, observers have identified amongst the causes of this awakening a set of common factors including social inequality, high levels of youth unemployment, a desire for greater individual freedoms together with a widespread sense of dissatisfaction in the population, stressing that at the root of the protests were significant factors unrelated to income.[3] This having been said, one cannot underestimate the impact, especially in some countries, of the increased weight of food expenditure on family incomes brought about by the volatility in prices of agricultural commodities since 2007. The reduction in access to food caused by rising prices was a co-factor, the spark that has undermined the foundations of the 'triangle of oppression'.[4] On the other hand, we are talking about a phenomenon that has always accompanied the history of rebellions, movements that led to the overthrow of long-standing regimes, some having lasted centuries. When dissatisfaction spreads on an empty stomach it becomes infectious and what starts out as a simple uprising can become a revolution. In France in the eighteenth century, a worker spent on average 50 per cent of his/her wage on the purchase of bread, a figure which had grown close to 88 per cent by 1789, the year of the storming of the Bastille. In Russia in the first months of 1917, the collapse of the supply system caused by the country's participation in World War I, and the resulting increases in the price of food staples such as bread (+15 per cent) and potatoes (+25 per cent), triggered protests that brought an end to the reign of the Romanov dynasty, thus paving the way for the October Revolution. Recently, the New England Complex Systems Institute published a study analyzing the timing of the riots in North Africa and Middle East with regard to price spikes in 2007/2008 and 2010/2011, showing a striking coincidence in their dates.[5]

In December 2010, following the spike in prices of agricultural commodities, the global prices of sugar and cereals had increased by 77 and 40 per cent over the previous year. This was a decisive event for the stability of many Arab countries, where the availability of the two commodities, which provide 61 per cent of per capita calorie intake in these regions, is largely dependent on imports, with 58 per cent of cereals and 75 per cent of sugar coming from abroad. This applies to a part of the world where the poorest of the population spend on average 65 per cent of their income on the purchase of food.[6] In setting out these figures it is not difficult to understand why the 'awakening' started in the food markets. During the first pro-democracy rallies in December 2010, the Tunisian protesters brandished *baguettes*. A few weeks later a surge of 20–30 per cent in the prices of staple foods in Algeria triggered the 'cous-cous revolt', protesters in Jordan and Yemen shouted 'bread and freedom' but it is Egypt that most clearly shows the link between the new world disorder, food supply and Arab awakening.

Indeed, Egypt represented the democracy of bread *par excellence*. The definition indicates a model of a tacit social contract in which, in exchange for low cost or free essential services such as education, employment insurance and health care, a highly authoritarian (or, at best, paternalistic) power is accepted by those it governs. The main symbol of this social contract was the system of subsidies that guaranteed bread at an affordable price, a system that began to appear very inefficient by the late 1970s, but at least represented a guarantee of social peace. When local governments attempted to abolish the subsidies during the 1970s and 1980s, bread riots multiplied across the region. Later, during the 2008 crisis, the rush for bread at low prices caused several casualties in Cairo, forcing the then Egyptian President Mubarak to militarise its distribution and production. The subsidy system was again plunged into crisis following the most recent price peak in 2010, this time bringing down with it regimes that had lasted decades, including that of Mubarak himself and the Tunisian Ben Ali.[7]

After the bread riots of the late 1970s, which caused 160 deaths in Cairo alone, Egyptian governments abandoned all thought of reform and began to lavish subsidies, basically flour at a reduced price for the bread-ovens, in order to keep down the final price of the product. Jordan and Yemen did the same, as well as Iraq under Saddam Hussein. Moreover, in the era of abundance the cheap raw materials necessary were not lacking, given the surplus being produced by western countries. The system, however, had hidden costs that would become obvious in the long run, such as the ongoing abandonment of domestic agriculture and the rise in youth unemployment against a backdrop of powerful population growth and rapid urbanisation. So Egypt, which in 1960 produced enough grain to feed its population, by 2010 was among the world's largest importers of wheat, and, whilst subsidising bread to the tune of about 3 billion USD a year, was unable to eradicate scourges such as that of child malnutrition.[8]

The grave instability brought on by the food crisis of 2007–2008 and still ongoing finally caused the system to crash. In 2010, in the face of the second sharp increase in grain prices in just over three years, the response of governments has done nothing but exacerbate the crisis. Egypt, Jordan and Yemen have increased subsidies, Assad's Syria withdrew its proposals to cut them, whilst Algeria, Tunisia and Morocco have cut import duties to zero; initiatives that instead of curbing rising prices have increased global instability, thereby worsening the situation. In the meantime, the revolt was underway, affecting the region, with bread as the trigger, a pretext, a terse symbol. Together with their subsidies, the regimes that had used them to define their relationships with their citizens went into crisis themselves. The Arab awakening is not limited to a loaf of bread, but it represents a particular case in which waving one has come to represent not only a desire for a better standard of living but a different way of thinking about participation and democracy as well.

A bitter awakening

On 25 January 2011, at the height of the surge in prices, Paul Kelly, Director of Food & Drink Industry Ireland (FDII), expressed his concern with regard to the security of the sugar supply within the European Union, drawing attention to the lack of the product available to honour existing contracts. Six months later, Dietmar Kendziur, president of the German Confectionery Industries Association, in the wake of fears surrounding sugar scarcity, expressed his dissent by stating that it was 'unacceptable that in the EU there is enough sugar to produce bioethanol, but not for food'.[9] The complaints voiced by the European industry have since been summed up in the statement by Mella Frewen, Director General of the CIAA (Confederation of the Food and Drink Industries of the EU) during the Ministerial Summit of the G20, 21 June 2011: ' We look forward to the outcome of the G20 Summit of Agriculture Ministers in Paris, where we hope that an Action Plan is agreed in order to provide short, medium and long-term solutions to problems currently faced by food manufacturers on the global agricultural commodities market.'[10]

So it was that in 2011, the food industry itself discovered the effects of scarcity, even asking the G20 to take action to stabilise prices and supplies. Still more explicit was Peter Brabeck-Lemathe, the president of Nestlé who, in an interview with The Wall Street Journal, spoke widely about food security, attacking, head down, the 'wrong policy-making' and its effect on 'millions of people, who we have been pushing back into extreme poverty' and the indifference of the west when faced with the issues of food supply, incentives for biofuels, fear of innovation and new technologies.[11] It is no coincidence that the first to move were the big companies with interests in sugar. On 29 December 2010, just days before the start of the new year, the futures traded for refined sugar on the LIFFE (London International Financial Futures and Options Exchange) in London touched their highest level in the last thirty years, passing $826 a ton. Just a few days before, the raw sugar futures quoted on the New York Board of Trade reached new highs, standing at around 34 cents a pound (about $755 a ton). What could justify such high price levels for agricultural commodities whose world prices throughout the 1980s and until 2005 consistently remained (apart from the odd peaks) under $300 a ton? The rise which sugar prices have undergone since 2009 is impressive, as Figure 4.2 shows.

The causes of this price surge are partly those that have pushed up the prices of other commodities. However, in the case of sugar other factors are at play that have destabilised the balance in world markets, especially the reform of the European system regulating the sugar-beet market, established in 1968 with the dual aim of ensuring a fair income for producers together with an adequate degree of self-produced supply for a commodity considered strategic. To this end, an intervention price for sugar was established that would be able to ensure the fair prices for producers of beet or cane sugar, a price which, over the years, has always been significantly higher than the world average.

The system was reformed in 2005 to meet the requirements of the liberalisation of trade, especially those from developing countries, which include those which are net exporters of sugar. With the reform the scale of public intervention was significantly reduced, leading to rapid downsizing in the European sugar industry production capacity. From over 22 million tons in 2000/2001, the EU's sugar production fell to just under 15 million tons in 2009/2010, against a domestic consumption of more than 16 million tons a year, creating a sugar deficit in the EU market, which has now become structural. The European Union has switched from being the world's second largest exporter of sugar (after Brazil) to being a net importer. The significant decrease in EU production has caused shortages in stocks of this foodstuff worldwide with negative economic conditions triggering repeated price increases in the commodity. In this way even food businesses have found themselves having to face a genuine sugar shortage on the market at very short notice.

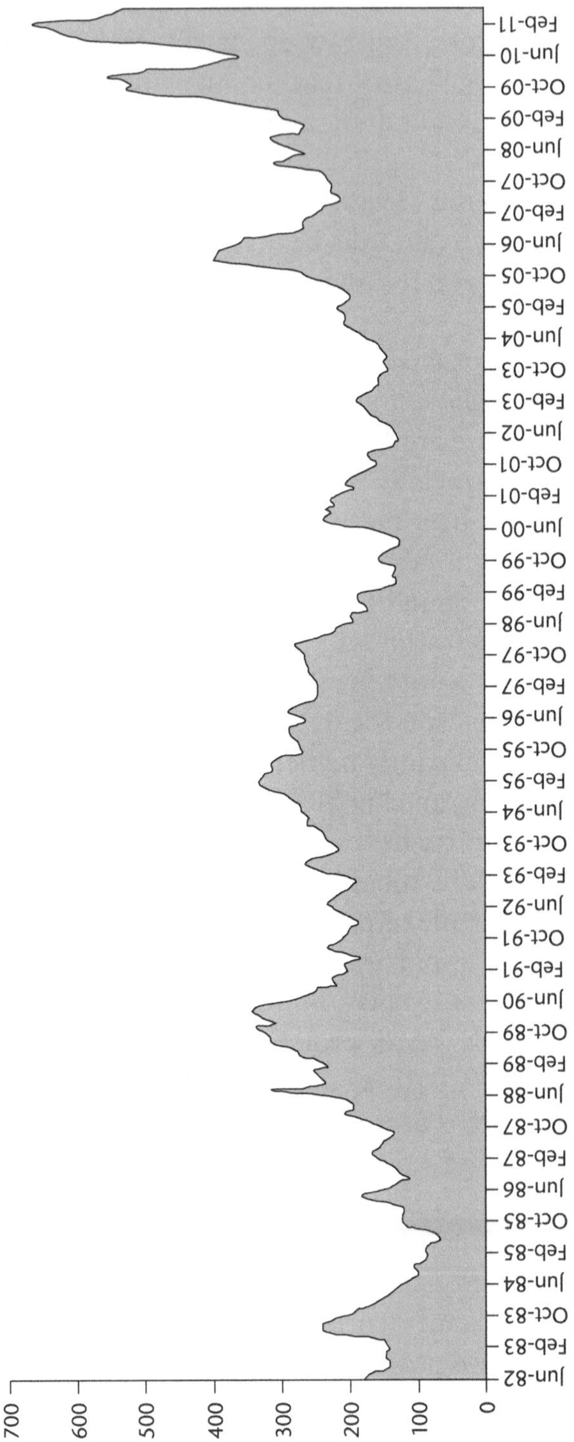

FIGURE 4.2 Sugar quotations week-on-week 1982–2010
Source: IMF

Our intention, however, is not to criticise the reform of the European sugar-beet industry, as it is entirely consistent with the aim of liberalising and streamlining the functioning of markets, but only to highlight how rapid and intense the change we are seeing has been. Until a few years ago, the selfsame European manufacturers who today are expressing concern over the shortage of sugar were shouting at the tops of their voices for the need to eliminate the guarantees granted by the EU to encourage a reduction in the price of this raw material, issuing statements of content completely at odds with those we hear from them today. 'We need a system of supply at competitive prices. This can only happen if European beet production becomes sustainable and competitive. We believe that the proposed reform can help to make it so', said Alain Beaumont and David Zimmer of CIUS (the Committee of Industrial Users of Sugar) on the eve of the reform. A representative of a small company producing biscuits added: 'The reform of the sugar CMO will allow Desobry to become a European company that is more competitive in exports. Sugar represents 21 per cent of all raw materials used, the cost of which accounts for 5.3 per cent of turnover. At these prices it is very difficult if not impossible, to obtain a half-decent return on investment. We need reform to become more competitive.' All these statements were made in 2005 by European sugar users (mostly in confectionery and bakery products businesses) and by some of their most important members after the presentation of the proposed reform of the Common Market Organisation for sugar by the European Commission.[12]

Today that scenario is pretty well reversed, and sugar is just the most sensational example of the situation affecting the food industry. Other products such as cocoa[13] and coffee are experiencing exceptional price rises affecting large parts of the 'colonial' productive fabric. The decision of the multinational coffee buyers to 'skip a step' in the distribution chain, contacting the producers directly instead of dealing with small farmer cooperatives, has triggered what has been called by some the 'War of the Beans'.[14] The greater purchasing power of these large traders compared to the slowness with which payments are made by cooperatives is leading many farmers to withdraw from the agreements to sell directly to corporations. The result is the closure of many cooperatives that are active in the harvesting and trade in coffee and the risk that looms over the long term for producers is to no longer have the economic and organisational structures that can support them on the market, with an impact that could prove devastating to local communities.

Although in a completely different manner, even in industrialised countries the relationship between industry and agriculture is changing. In the face of volatility and the concentrations seen in certain markets, concerns about possible breaks in supply are pushing processors and producers to take joint action to reduce risks. More and more frequently, they are entering into long-term supply accords based on agreed prices, or that, at the very least, vary within a predefined range. In some cases, now common within Europe, public policy-makers have become the promoters and guarantors of such agreements. This radically changes the situation that was well established in the food processing world of looking at the short term and buying at the lowest price on ebbing international markets. As a consequence of the transition to an era of scarcity, the food industry must now also insure its most distant horizons to assure the economic viability of its manufacturing base.

5

LAND AHOY!

The deal of the century

'It's very good land. It's quite cheap. In fact it is very cheap. We have no land like this in India.'[1] When the project manager of Karuturi Global Ltd., in the region of Gambella, in western Ethiopia, was talking to the Guardian reporter John Vidal, he probably couldn't believe his luck. According to the picture painted by the journalist, the agreement that the company from Bangalore, the world leader in the production of roses, had signed with the government of Addis Ababa covered the rent of 300,000 hectares of fertile land for around 200 USD a week to be given over to the production of rice, palm oil, sugar cane and cereals for a period of fifty years. According to the investigation of another journalist, Stefano Liberti, Karuturi will not pay rent for the first six years, and will then pay less than 1 USD a hectare for the next 84.[2] Different versions of a private agreement, on which there has never been any public discussion. In any case, the price is rock bottom. 'It's the deal of the century,' wrote Vidal in March 2011. But for whom?

For Karuturi, the long-term lease of large swathes of arable land represents a leap from its leadership in floriculture into the company of the global giants of integrated agricultural production. The Bangalore-based company, which mainly produces roses for the European market, has been present for years in Africa, but 'the acquisition of large tracts of land in Ethiopia – one can read on the company website – has set the stage for us to become a complete agriculture production company. Our goal now is to make a significant contribution to alleviate the global and African food crisis.'[3] In Gambella, Karuturi has already invested over 100 million USD and promises to bring jobs and infrastructure, new technologies and an efficient model for the production of foodstuffs intended primarily for the local market, building roads and bringing in essential services and building, close to what promises to be one of the largest farms in Africa, a city of 15,000 inhabitants.

For the Ethiopian government the agreement is synonymous with modernisation. After decades of aid that proved unable to rescue millions of people from poverty and malnutrition, in the view of Addis Ababa foreign direct investment in local agriculture and its transformation into a large-scale industry represents the hope leaving behind subsistence farming, the spectre of underdevelopment and the reality of hunger, which have made the country the very symbol of the inability of most of Africa to feed itself. It remains to be seen whether the only way to achieve this development is to sell off the country's land. In Ethiopia, according to some sources, the supply of virgin land leased in recent years now covers more than 3 million hectares, all areas that the government says are 'sparsely populated'.

On paper, the deal seems advantageous, for the company and for the development of the country itself, but many parts of the deal remain unclear. Vidal recounts the reality of deforestation, the forced clearance of fifteen thousand people who lived in the area. 'A coincidence' explain the government sources in Addis Ababa, an initiative that was independent of the arrival of Karuturi in the region and part of a relocation programme aimed at improving the living conditions of communities. Then there are the local workers, who, says the British journalist, are being paid wages below the poverty line. What is more, according to other sources, farmers from the

Punjab are signing memoranda of intent with Karuturi allowing them to manage large parts of the area directly.[4] The deal of the century seems to exclude the interests of those that both the Ethiopian government and the Indian company itself are claiming to be its primary beneficiaries.

In May 2011, the government in Addis Ababa announced that it had, in agreement with the company, curtailed the lease to Karuturi by two-thirds, scaling it back from 300,000 to 100,000 hectares. The company replied by denying the fact, calling the news 'completely baseless'.[5] The company says it will continue to work on the 100,000 hectares with the rest being used in a second phase, as planned all along. The official reasons for the reduction in the area covered by the agreement given by the government in Addis Ababa concern the contract, which was apparently signed without setting strict boundaries and precise geographic coordinates, the size of the area being too large to be managed by a single company, and the impossibility for private management to ensure the right conditions for the annual migration of antelope in one part of the land being leased. The contract, traditionally an element of guarantee for the parties, has ended up being the most opaque and uncertain part of the whole affair, not only regarding its effects upon the lives of the region's population, but for the very people who have undersigned the contract.

The Gambella agreement is just one of many announced or concluded deals made in recent years by private companies, governments and sovereign wealth funds around the world, for the acquisition of land on a large scale for the production of staple foods and biofuels in developing countries. It represents a phenomenon that has accelerated since 2007/2008 to the extent that it is being followed with interest and concern by intergovernmental institutions such as the World Bank, non-governmental organisations dealing with human rights and nation states such as Japan, which, referring to the emergence of the phenomenon, wanted to put the theme of 'responsible investment in agriculture' at the international level on the agenda of the 2009 G8 Summit. Some NGOs and the media from around the world call it 'land-grabbing', the hoarding of land. Despite receiving a lot of news coverage, its precise scale is as yet unknown. The indisputable fact is that there are many people who are willing to shell out billions for the right to use large areas of arable land, often only nominally 'virgin', marginal or depopulated, and that some people are very willing to do a deal on these areas. The deal of the century? It's the Earth itself.

BOX 5.1 NO MAN'S LAND, BY TRADITION

One of the drivers of large-scale acquisitions of land is the tendency to consider as unproductive, empty or abandoned any areas that actually provide a livelihood for thousands of families. These are small farmers, pastoralists, communities that use these fields and pastures by custom and that cannot exhibit title deeds even if they have cultivated them for generations and which are essential for their survival. If this informal title by use on pieces of land is not recognised or formalised by the state, they are virtually unprotected in situations in which this land is being put to another use or sold off, and are forced to leave areas where they have always lived without consultation or compensation. This is a situation that is more common than you might think. According to estimates from 2005 the vast majority of households (76.1 per cent) who live in the rural areas in developing countries have no title deeds,[6] a situation that often makes areas where access to land is vital to survival endemically unstable and the right to exploit it becomes a source of recurrent conflicts.

Soil, the last refuge

The rush to grab agricultural land, triggered by the 2007/2008 food crisis, is a fact that says more than any other about the spirit of the age of scarcity. Sovereign wealth funds, giant global farms, investors with a good nose, financial companies in Europe, Asia and America are buying or taking out long-term rents on large areas in Latin America, Asia and the former Soviet Union, but especially in sub-Saharan Africa. Some call this phenomenon 'land-grabbing', linking it explicitly to colonialism. Others prefer to speak of 'large land acquisitions'.[7] Others still use the general term of 'land rush', or the more neutral 'land deals'.[8] These are nuances that are often synonymous with opposing viewpoints in their evaluation of the phenomenon and the actions to be taken to regulate it (see Box 5.2). However it is defined, the demand for land has acquired extraordinary dimensions in the space of a few years. According to the latest estimates, since 2000, 50 to 80 million hectares have been the object of negotiations around the world, more than two-thirds of which is in sub-Saharan Africa, with Ethiopia, Mozambique and Sudan having the largest areas. With 45 million hectares being negotiated between 2008 and 2009, one can see that the phenomenon is undergoing a major acceleration, given that until 2008 the expansion of global arable land proceeded at an average of 4 million hectares per year. Of course, these figures do not differentiate between rental and purchase of land or between agreements that have been announced, signed or are actually in operation. This is important to underline given that only 21 per cent of the contracts reported go on to be successful. According to other sources, however, 20 million hectares exhibiting the characteristics of large-scale land acquisition were actually exchanged between 2006 and 2009, an area larger than Senegal.[9]

In the executive phase of this type of transaction, usually characterised by a low level of transparency, there are a number of recurring problems including a lack of consultation with local communities and social and environmental impact assessments, insufficient consideration of the importance of ecosystem equilibria and the sustainable management of land, water and biodiversity along with the production of food intended primarily for export from areas known for food emergencies, a situation which is aggravated by the strong tendency to substitute food crops for energy crops. At the same time, it is just this type of agreement that marks a sharp increase in direct investment in agriculture in developing countries which, if properly directed, could be exploited to trigger rapid growth in agricultural production, contributing to the stability of local and global food supplies.

BOX 5.2 WHAT'S IN A NAME? 'GLOBAL LAND-GRAB'

The commonly used expression 'land-grabbing' historically referred to colonial practices and conflicts relating to the expansion movement of the frontier in the United States. For several years it has been used by some non-governmental organisations to indicate the acceleration of the large-scale land transactions triggered by the food crisis of 2006–2008. According to former Secretary-General of the FAO Jacques Diouf, the new 'land-grab' presupposes the risk of a 'neo-colonial pact for the supply of raw materials'.[10] Tying the phenomenon to the meta-narrative of colonialism has contributed to its high newsworthiness and, consequently, its claim to a position high up the global media agenda, to the extent that some reports speak of the 'land-grab hype', fearing the risk that the high threshold of attention towards the phenomenon is accompanied by an equally high degree of confusion.[11] The term 'global land-grab' has become an umbrella term to describe and analyse the current explosion in transnational commercial land transactions but it does not help to define its real boundaries. The neo-colonial model cannot explain the more general trend to increased demand for land at every latitude, or the movement towards the concentration of land ownership at the regional level, or even investments in arable land in the countries which are net food exporters with strong political stability. Although the issue has attracted a large number of publications – made up of news items, grey literature and the reports of inter- and non-governmental organisations – despite everyone agreeing on its importance, it not only lacks agreed monitoring systems and a transparent benchmarking operation, but even an adequate analysis of its impact in the medium and long term. Indeed, the data that would contribute to a more rigorous definition is lacking, or, worse still, conflicting, from the point of view that takes in the description and analysis of the phenomenon.

There are two positions that animate the debate, with many gradations between them.[12] There are those who, like most NGOs, approach the phenomenon from a human rights and social sustainability perspective. A perspective that, based on the information that these organisations collect in countries that are spearheading the supply of land, legitimises the use of the term 'land-grab' and the neo-colonial narrative that land acquisitions are essentially a threat to life and livelihoods of the rural poor, the balance of ecosystems and regional and global food security, and should be severely restricted, if not banned.[13] Others prefer to speak more generally of 'land deals'. The International Land Coalition, an umbrella group which includes intergovernmental organisations and civil society, uses the term 'commercial pressures on land', extending the field of analysis of land investment to those originating from the tourism industry or from the trade in carbon credits.[14] Also to be taken into account is the importance of acquisitions in the context of international investment in agriculture in developing countries, that is to say a prospect of economic sustainability without which it is difficult to imagine valid paths of social sustainability.[15] Some observers of intergovernmental institutions, while placing the emphasis on the risks of such transactions, especially in unstable areas of the planet, see in them an opportunity for economic growth and development, linked to the flow of private capital which, if properly managed, could promote a rapid increase in agricultural productivity in these areas, forming the basis for stable and lasting growth. The World Bank prefers to describe the phenomenon by using the phrase 'large-scale acquisition of agricultural land'. Some even more explicitly invite the 'making a virtue of necessity' to 'address the threats' and 'capitalise on the opportunities' connected to direct foreign investment in the acquisition of land.

From whatever perspective we analyse the phenomenon, however, the creation of shared information infrastructure seems ever more necessary to provide the essential information required for its monitoring and to tackle critical issues. The land rush does not seem destined to stop, and the case-histories collected thus far by the vast majority of sources highlight a few virtuous examples among many agreements that do not seem to bring benefits to local people.[16] In short, 'land deals' are few and far between, and there are too many 'land-grabs'.

The increased demand for arable land is real and demonstrated by the rise in prices on a global scale. In 2007 alone the value of farmland increased by 16 per cent in Brazil and 31 per cent in Poland, while in the USA there has been a increase in value of 20–70 per cent in the last five years, depending on the land's geographical location and suitability for use in intensive production.[17] The land rush is a direct manifestation of the uncertainty that dominates the global food supply system. We find ourselves in a phase of population growth with increased demand for food and low levels of agricultural stocks, coinciding with a progressive degradation of natural resources that threatens future food production capacity, together with the feared effects of climate change on the production of food, the malfunction in global and regional financial markets and the failure of distribution policies and development assistance. Taken together they mean that states are seeking new means to produce food and large private companies smell a profitable and long-term business opportunity. The competition for land use incentivised by policies to promote first-generation biofuels has only exacerbated the situation. If, as we have seen, land turns out to be a scarce resource, the rights to its exploitation, including the water it contains, is a strategic factor crucial in any future geopolitical balance.

It should also be pointed out that the appropriation of extensive areas of arable land in other countries and continents by large private companies is nothing new. The history of colonialism is full of such episodes in the nineteenth and, later, at the beginning of the twentieth century. What is characteristic of the new global land-grab is the intensity of the demand, the extent of the areas involved and the negotiation of a formal contract, although in most cases this proves incapable of providing real guarantees to the contractors or social security for the communities affected and differs in that the appropriation is completed on paper and not as a result of the persuasive force of 'the barrel of a gun'.[18] Another difference with the past is the type of crops involved which are not directed towards cash crops such as cotton, rubber or sugar but towards staple crops and biofuels.

A leading role is being played by sovereign wealth funds in the new generation of land deals, that is to say by the states that invest abroad to secure themselves channels for the supply of food, 'in the same way that oil investments are made by state enterprises and are designed to achieve energy autonomy for a nation'.[19] The governments most openly engaged in these types of negotiations are those of the former 'Third World', a situation that constitutes a pattern of relationships of a 'South–South' kind, although not applicable to all land deals, and represents another difference from the colonial practice, in which the pattern was essentially North–South. A paradigmatic example is that of the Gulf States, with very little arable land, large budgets and increasingly dependent on imports to meet their food needs. In 2003 the States of the Gulf Cooperation Council (GCC) spent 8 billion USD a year on food supply, but by 2010 the outlay amounted to 25.8 billion USD and the projected figure is expected to more than double in the coming decade to reach 53.1 billion USD by 2020.[20] With such an expensive shopping trolley, renting the land to produce the food directly seems logical, especially at the bargain prices on offer in Ethiopia or Liberia, where renting a hectare costs, on average, respectively, 2 and 5 USD a year.[21] The purchase prices are also competitive, just 800–1000 USD per hectare in sub-Saharan Africa as against 6,000 USD in Brazil, 18,000 USD in the United Kingdom and up to more than 22,000 USD in Germany.[22] It goes without saying that countries like the UAE, which in 2008 imported 85 per cent of its food, paying 2.9 billion USD, are very interested in negotiating, for example, a 400 million USD project for the long-term rent of 100,000 hectares in Pakistan; or that Saudi Arabia, after having launched, tested and abandoned a huge self-sufficiency project in 2009 should have created the King Abdullah's Initiative for Saudi Agricultural Investment Abroad (KAISAIA), with an initial budget of over 600 million USD with a mission to contribute to national food security; or that the Saudi Company for Agricultural Investment and Animal Production should have the resources for a capital of 800 million USD to invest in agricultural projects abroad.[23] The move is as logical as it is epochal. In the era of scarcity, instead of going to the market, where the merchandise is not abundant and ever more expensive, it has become cheaper to buy the asset directly to produce it.

Although one of the main novel elements, involving deals of considerable size in terms of the areas and sums of money involved, the presence of the financial arms of sovereign states in land deals is part of a broader context, in which government support is sometimes indirect. Probably the best known case of this kind, discovered by the *Financial Times*, in which the main player was not a sovereign wealth fund but a commercial firm, took place in November 2008 when, with the quote 'food can be a weapon in this world',[24] the South Korean multinational Daewoo Logistics announced it had reached an agreement with the government of Madagascar for the exclusive use of 1.3 million hectares (half of the country's cultivable area) to produce maize and palm oil for 99 years at no cost. The deal was torn up following a coup that led to the government by the Mayor of Antananarivo, Andry Rajoelina, with the backing of the army.

However, there are a wide range of competitors racing to get their hands on land, not just sovereign wealth funds, made up of a wide range of bodies. Apart from the companies involved in large scale agricultural production, there are private investors, pension funds, financial players searching for the last safe haven, the oldest and most secure of investments, land itself. But in the era of scarcity the investment itself ceases to be 'safe' or at least comfortable, especially where it is cheaper to buy at low prices, for example in the south, where any land deals go together with great uncertainty and an appetite for high risk. First, given the size of the areas involved, it is very difficult to manage such transactions, even for companies with some experience in large-scale agricultural production, let alone for investors from other sectors. Second, the potential for damage to reputations is very high because an acquisition policy, even if formulated as a long-term lease, may prove unsustainable without an acceptable level of satisfaction among local populations, which may lead to disputes over property rights,[25] and not only in areas where such rights are primarily ordained by custom and tradition. There are cases where, even if the process of acquisition by the investor appears sound, the weakness of local administration combined with a lack of clarity in formal or informal property regimes can lead to situations of conflict with small-scale farmers who end up losing land against their will. Pushed on by the high value of land, 'domestic land-grabbing' has also developed in which the average size of acquisitions is reduced, becoming the preserve of domestic investors, members of urban and military elites, actors of the states involved, a fact confirmed by the percentage of domestic investors on arable land in Cambodia (70 per cent), Ethiopia (49 per cent), Mozambique (53 per cent), Nigeria (97 per cent) and Sudan (78 per cent).[26] Rather than a 'land-grab' in the sense of colonialism there is a phenomenon of the excessive concentration of land holdings, the consequences of which may prove equally risky.

There are many, therefore, who want to dive into the 'deal of the century', and financial integration in the global economy does not help to simplify the interpretation of the phenomenon. For example, the important news that, within their investment portfolios, American universities own vast tracts of land in Africa, managed for them by hedge fund companies, aroused amazement.[27] What are the implications of participation in the land rush of pension funds, the performance of which will be crucial to ensuring a peaceful old age for millions of people at a time of a reduction in public spending on pensions in the developed world? The rush for land has taken on a dimension and a complexity so great as to overshadow the historical significance of another rush, that of 1849 for gold in California. Today, gold has become land and the competition for it has become global and the stakes are much higher for everyone.

Who's buying and who's selling

Why, despite the high risks for investment, do companies like Karuturi Global Ltd. put in place strategies for the acquisition of large areas of arable land thousands of miles away from where they are based? Why, in the context of stiff competition for the use of land and new challenges to the food supply does demand meet supply at reasonable prices, even if, as in this case, the supply comes from Ethiopia? This is a country that is symbolic of the food crisis and the failure of policies to support development that should have combated it, a country that in 2009 alone received nearly 4 billion USD in official aid and was recently hit by the famine in the eastern part of the Horn of Africa, the worst since 1950.[28] On the other hand, the south-eastern part of the continent represents a kind of hub of large-scale land acquisitions by foreign investors where Sudan, Ethiopia, Kenya, Tanzania and Mozambique lead the way in the size of the areas negotiated under formulas involving their long-term lease. In the west of the continent the cases of Liberia and Mali are to the fore, with agreements for the use of hundreds of thousands of hectares for agricultural and agro-industrial activities.[29]

If Karuturi behave as a company and seek to maximise their profit, why have some African governments treated their territory as a sort of global land supermarket? 'This apparent rush to lease so much farmland, much or all of it to foreign interests, during a time of worsening food insecurity should raise concern within the donor community about the G[overnment] o[f] E[thiopia's] motivations and negotiating capacity' said a cable from the US Embassy in Addis Ababa in December 2009.[30] Apart from considerations of effective representation and the ability of some governments, it seems that the African continent's elites regard foreign direct investment, even in the form of the sale of high quality land at laughable prices, as the only way forward for development. The problem for Africa is that despite billions in aid being granted over the decades,[31] the quality of investments has not provided local producers with access to the technical improvements which have benefited those in certain parts, say, of Asia. Within this framework it is the very structure of African agriculture itself that is being questioned by local elites, who give great weight to the investments of foreign corporations in the construction of farms that are large enough to be able to develop economies of scale. So, while intergovernmental agencies are seeking solutions to integrate small farmers in the southern hemisphere into the processes of development,[32] relocating and even moving by force the population from smallholdings, considered synonymous with low productivity, has become the policy of many African governments, even if it remains unstated.[33] In this cultural climate, the long-term renting out of pieces of land to sovereign wealth funds and foreign firms is seen as a quick fix, which brings in capital, guarantees the construction of infrastructure and lends impetus to the industrialisation of agricultural production.

Despite the progress made in Asia in the field of agricultural productivity, the south-east of the continent is among the main targets of investors, particularly (but not exclusively) Laos, Vietnam and Cambodia for their forest resources and the Philippines and Malaysia for the cultivation of oil palms. The arable land of Pakistan has effectively been on the market since 2009 when the then minister for privatisation in Islamabad told Reuters that he wanted to hand over 2.4 million hectares to foreigners.[34] Land in Brazil is particularly sought after in South America, or at least it was until the recent introduction of a rule that limits its purchase or lease by foreign investors, an example that Argentina may soon follow. In Europe, there are agreements, signed or entered into, covering land in Ukraine (especially from private investment funds) and Russia and even Australia and New Zealand are not immune from the interest of competitors in the land rush.

State intervention

The acquisitions, entered into or announced by sovereign funds, the states or by corporations supported by national governments, are often very sizeable in terms of the sums of money involved and the pieces of land acquired and are also those that receive the most coverage in the press, even when, in the end, they fail to go through. As already mentioned, the countries moving with greatest alacrity on this front are the Gulf States. Qatar proposed 3.5 billion USD funding for the Government of Kenya for the construction of the new port of Lamu, in exchange for the rent of 40,000 hectares for the production of fruit and vegetables in the Tana River Delta (an area in which the government in Nairobi is offering large areas of fertile land to domestic and international investors), but the negotiations broke down. Kuwait has rented (at a cost of 546 million USD for a period of 70 years) an area of between 50,000 and 130,000 hectares for the cultivation of rice in the province of Kampong Thom in Cambodia. The United Arab Emirates has bought

324,000 hectares of land in Pakistan and in the coming years plans to put into production 3,000 hectares of bananas in the Philippines and 30,000 hectares of cereals in Sudan. The case of Sudan is somewhat similar to that of Ethiopia. With an ongoing humanitarian emergency in Darfur where more than 2.5 million refugees are attempting to survive at this very moment, between 2004 and 2009 the Sudanese government transferred millions of hectares of land to foreign investors and 'in 2008, while the United Nations were forced to cut emergency rations because of the hike in food prices, Khartoum doubled its exports of cereals'.[35] Many are attracted by the Sudanese land market, especially the Gulf States, but Jordan and South Korea are also involved and Gaddafi's Libya was one of the most active players on the continent. The agreement in June 2009 between Malibya Agriculture and the Government of Mali witnessed the renting out (for 50 years with option to renew for up to 99 years) of 100,000 hectares of irrigated land in the Niger basin. The goal is to produce rice and other agricultural products to supply the needs of Tripoli but all this without any consultation with the area's 75,000 inhabitants and without conducting any assessment of the social and environmental impact whatsoever. So far the local population has not received any benefit from the agreement and it remains to be seen what will become of it after the fall of the Gaddafi regime.[36]

India definitely has its sights on East Africa. In August 2011, a delegation of 35 firms willing to invest up to 2.5 billion USD visited Ethiopia, Uganda and Tanzania, where Karuturi Global Ltd., has announced plans to invest 500 million USD for oil palm cultivation, sugar cane and cereals on an area covering 350,000 hectares to be acquired on a lease. In the delegation were also McLeod Russel India Ltd., the world's largest producer of tea, the seed company Kaveri Seeds and the Escorts Group, mainly involved in the manufacture and sale of agricultural machinery. New Delhi supports such moves not through sovereign wealth funds, or direct government investment, but by funding local projects through credit, preparing the ground for domestic enterprises to work more effectively. The project promoted by Tendhao Sugar, a refinery with a production capacity of 300,000 tonnes of sugar per year, promoted by the government in Addis Ababa, received a loan of 640 million USD from state-owned Indian banks. Work on the plant, which also requires the construction of an irrigation system for the cultivation of the raw materials, started in 2004 but has not yet been completed. Tanzania could benefit from more than 90 million USD of credit once it has paid back a loan of 40 million USD to another Indian bank which it employed to buy tractors from the Indian Escorts Group whilst farmers' organisations may soon arrive via the open channel of credit and the large corporations. Under the plans of the Confederation of Indian Seed Potato Growers (Poscon) 2,000–5,000 hectares would be leased in Ethiopia for the production of oilseed, pulse, sugar cane and maize.[37]

India's main commercial competitor in East Africa is China which, while being a major player in foreign investment, does not seem to show any great interest in large-scale acquisitions. Recently, a entrepreneur from China came forward to buy 30,000 hectares on which to build a tourist establishment in Iceland, but the acquisition of arable land, especially in Africa, was explicitly excluded from the twenty-year strategy on food security published by Beijing in 2008, with the possible exception of land for the cultivation of soybeans in Brazil. The 'cooperation farms', Chinese-owned quasi-state organisations in Africa, are usually less than 1000 hectares in size, while many land negotiations announced in previous years have not been finalised. China is still engaged in a number of agricultural projects abroad but rarely with a formula that involves the direct commitment of public money. More often it uses a system of multinational corporations that move with the support of the central government or within a framework reflecting the national interest. These are projects that not only target Africa, where for years Beijing has had a deeply entrenched trade network offering investments in infrastructure in exchange for raw materials,[38] but involve the rest of Asia, the former Soviet Union, Latin America and Oceania.

For more than ten years, China has promoted joint ventures with Cuba and Mexico for the production of rice. More recently it has proposed the stipulation of several agreements in Kazakhstan to produce food for the local market and worldwide. In this case an agreement for the purchase of thousands of hectares was denied by officials representing the former Soviet republic. Less clear elements surround the exploitation of Cambodia's forests, which began in 2007 in a country with weak land ownership governance that is also under the influence of the interests of firms from South Korea and Bahrain. Commercial relations with Brazil are of particular importance which, in recent years, China has tried to make more stable, partly to diversify a portfolio investment that has become exposed to US debt and partly to secure a privileged channel for the food that can be supplied by the largest agricultural power in the world.[39] One of the most impressive deals in terms of the volume of agricultural commodities and money potentially involved is the ongoing negotiations with the State of Goiás for an investment of 7.5 billion USD, which aims to begin production of 2.5 million hectares of soybean, of which 6 million tons a year will be allocated to China. This is not a purchase of land, but an investment in the Brazilian state's transport infrastructure and the supply of pesticides, seeds and fertilisers, in exchange for part of the harvest, this being one way to ensure production without violating the aforementioned limitations imposed on the possession of land by foreigners laid down by Brasilia in 2010. The order, signed by former President Lula, established ceilings of between 250 to 5000 hectares for land ownership by foreign investors depending on the region, though this, however, is to be subject to revision. President Rousseff's Executive is considering the possibility of allowing long-term rental, given that the initiative which places limits on land ownership has cost 15 billion USD in lost investment in a single year.[40] In Argentina as well, President Kirchner has proposed legislation similar to that of Brazil, which would limit land purchases by foreigners to areas covering a maximum of 1,000 hectares and which do not exceed 20 per cent of the total national stock of productive farmland. In the meantime the Chinese Beidahuang agricultural corporation has secured the exclusive right to the soybeans grown on over 300,000 hectares for a period of twenty years in the province of Rio Negro, an area not in production at the time of the agreement, in exchange for investments in infrastructure, including irrigation channels. The project is controversial, triggering protests from the local population.

When it comes to increasing agricultural production, China itself is not closed to foreign capital. As a result of a joint venture between companies in Beijing and the government investment agency in Singapore (International Enterprise, I.E.), already active in acquiring land in Laos and Cambodia, there are plans to create a 145,000 hectare farm in north-east China that will produce food for the domestic and international marketplace.[41] Then there is the Oceania 'front'. In Australia, the wave of Chinese immigration originating from the 1867 Gold Rush resulted in the permanent presence of a strong community in Queensland, which saw the first farms being set up in 1989, well ahead of the 'land-grab hype'.[42] In the same state, in June 2011, the Chinese company COFCO formalised the purchase of a majority stake in the Tully Sugar refinery, for decades the sole property of the local sugar cane farmers. In the western part of the country Beidahuang intends to acquire 80,000 hectares. With intervention from their ambassador, Beijing began to pressure the Government of New Zealand to get the green light for investment in land and farms that would make the country the world's largest exporter of milk. The goal would be to place Fonterra, a dairy multinational owned by 13,000 New Zealand farmers, in a dominant position in China's internal market. In spite of concerns about an alleged Chinese invasion, however, data from the Overseas Investment Office shows the purchase of 170,000 hectares of land sold in New Zealand over the past six years were mostly bought by American and European (especially

German) companies. What if the government's investment in the acquisition of farmland, despite the great noise surrounding it, were just the smallest slice of the pie?

Private initiatives

Sixty million hectares, a little less than the surface area of France: this would seem to be the magnitude of the 'land-grab' by private investors in Africa alone, and this in a single year, 2009, according to the California think-tank, the Oakland Institute.[43] These are figures which, if confirmed, would prove that the operations supported by governments are basically small compared to the large-scale land acquisitions carried out by private bodies. However, given the institutional nature of such transactions, they are usually accorded less publicity than arrangements between states, and their potential outcomes or benefits to any contractors or interested parties are, if anything, even less clear. In this segment of the land market, moreover, the boundary between the transactions that are announced and those that are actually implemented is even more difficult to pin down even though it is of great importance in the identification of any speculative trends.

The main private investors are a varied group. There are those such as the European banks that have guaranteed funds for investment in oil palm plantations in Malaysia and the Philippines or have committed capital to energy crops in developing countries that have become economically profitable as a result of the policies and incentives and the system of carbon credits.[44] For some years now, prominent figures in the world of high finance have begun to invest in farmland, including businessmen such as George Soros, who controls 23 per cent of Adecoagro SA, a venture specialising in agricultural lands in South America. In 2008 Lord Jacob Rothschild, a descendant of the dynasty of bankers, invested in Agrifirma Brazil, the buyer of 40,000 hectares with options for another 24,000 in the State of Bahia. Another shareholder in Agrifirma is Jim Rogers, the prophet of high finance who, in the 1990s, had already foreseen the great deals to be made in shares involving raw materials. Beyond this, there is only the Earth itself: 'Everyone said, "Buy gold"' Stephen Diggle, a hedge fund manager and purchaser of land in Uruguay, the USA and New Zealand, told the press, 'but at the end of the day you can't eat it. If everything else goes and I just have these farms, it makes me moderately wealthy.'[45]

For several years investment funds have begun to offer specific solutions for agriculture, including land acquisition. Among the players in the financial sector there are those who attach more importance to the return on investment and those who prefer stability. The former are more risk-oriented, perhaps acquiring assets where prices are lower but where conditions for cutting the deal are more uncertain, as is the case in sub-Saharan Africa; the latter place more stress on stability and long-term returns on the investment and are interested in purchases that are perhaps more costly, but located in more politically stable areas of the planet. The first group includes companies such as Emergent Asset Management (EAM), based in the United Kingdom, with interests in land in sub-Saharan Africa, where it intends to acquire more than 50,000 hectares for the purpose of producing agricultural commodities for the local market. The second group includes pension funds, and according to the non-profit organisation GRAIN, of the estimated 100 billion USD of funds invested in commodities, 15 billion USD have been used in the acquisition of agricultural land.[46] The investors include the American fund TIAA-CREF, which has committed 2 billion USD in fertile land and farms in North and South America, Australia and Northern Europe, the Dutch pension fund APG (1 billion EUR), the American Ascension Health, (1.1 billion USD) and AP2, that is to say the Second Swedish National Pension Fund, which in 2010 began to invest in real estate and forest land in America, Australia and New Zealand with 500 million USD. 'Both forest and agricultural real estate are deemed to generate a stable return,' says the Swedish fund's 2010 Annual Report 'given that the underlying trend towards increased urbanisation and continued economic progress will increase the demand for forest and agricultural products.'[47] AP2, TIAA-CREF and other pension funds adhere to the UN Principles for Responsible Investment, and have promoted a code of self-regulation on investment in land devoted to agricultural production. As institutions required to pay pension benefits, i.e. a service to be provided in the long run, in general they invest with an aversion to risk. In this respect land is no exception, and is focused primarily on commitments in countries that are net exporters of foodstuffs and that are characterised by strong political stability, but where it is more difficult and more expensive to find areas that can be acquired. According to Jose Minaya of TIAA-CREF the arrival of institutional investors in land deals helps to increase transparency and promote good practices, including environmental sustainability.[48]

NGOs consider financial initiatives on land transactions rather dangerous. The Oakland Institute accuses the agricultural funds such as the one created by EAM of creating the conditions for the continuing uncertainty and insecurity in global food supply, constituting a 'bigger threat to global security than terrorism'. In its report on Africa, this activist group tells of deals sometimes concluded 'with a bottle of Johnnie Walker' and agreements on paper that promise to improve the living conditions of the population, but instead are responsible for the 'displacement of small farmers, environmental devastation, water loss and further political instability'.[49] The EAM has already called the allegations 'grossly inaccurate', stressing the positive impact of its investments in Africa in terms of employment and increased income for local communities.[50]

Who is right? It is hard to carry out a real cost-benefit analysis when there are no independent monitoring agencies, much less consolidated data and when the versions of the investor, the state authorities and the inhabitants of the area diverge in an almost systematic fashion. The same happens with foreign land acquisitions made for the purpose of direct production of foodstuffs. Entrepreneurs such as Calvin Burgess of the American Dominion Farms, the owner of thousands of hectares in the Yala swamp in Kenya for the production of rice (with associated tanks for aquaculture) and with the acquisition of a further 30,000 hectares underway, are convinced that they represent the incarnation of the pioneer, who have relocated to rural Africa for economic and ethical reasons and often come into conflict with local communities. For the Kenyan government, however, if there are problems, these are private.[51] It is just this absence of clearing-houses, consultation and exchange between stakeholders which represent the most critical factors in the land rush. Rather than defining investors as 'land-grabbers' outright, triggering an endless battle between farmers' rights supporters on the one hand and investment on the other, it might be more useful to focus on governance in land matters, the structural weaknesses of which in some parts of the planet may well give rise to much wider conflicts than those confined to the local sphere.

An 'unconventional factor of insecurity'

Recently, Saturnino Borras, an associate professor at the Institute of Social Studies in The Hague and one of the most active researchers in the field of the 'global land-grab', said he was contacted by the Dutch Ministry of Defence to discuss 'unconventional security factors and risks'. 'Not much research has been done but various national and transnational global authorities are showing interest in the potential effects of the global land-grab in terms of security.'[52] He continues:

> By way of example, when the Libyans acquire hundreds of thousands of hectares in Mali it creates a situation in which foreigners own a piece of territory of another state. To make everything more disturbing still, these are agreements that even if the formula is that of a lease, it lasts 99 years

> and is renewable. A very long time, implying management by different governments and different generations. In other African countries, China is sending a large number of workers from the motherland, that remain in the area as permanent migrants. Russia has recently returned to Ukraine, not as an occupying force, but to buy land. All these movements are of interest to the Dutch security authorities, but also to those in South Africa and multilateral organisations such as NATO, bodies that usually consider issues related to agriculture, food or biofuel as only marginal.

The interest of NATO – which has established a Public Health and Food/Water Group – is real, but on a more general scale. The experts of the Atlantic alliance have not directly and specifically tackled the phenomenon of land transactions as the factor that might affect food security in certain parts of the planet but 'amongst these – confirmed by NATO headquarters in Brussels – there are the sharp increases in food prices, the role of biofuels and, potentially, land-grabbing as well.'

The impact of land deals on the political stability of areas where access to fertile land and water is vital for millions of individuals is therefore a risk to take into consideration. On the other hand, if the risk that large-scale acquisitions of agricultural land could unleash conflict has not yet materialised, access to natural resources is one of the oldest triggers for armed conflict. The rush for land and land transactions on a large scale may be factors in 'unconventional' insecurity but they are linked to well-known dynamics. Take the case of the conflict in Darfur, Sudan. Although its causes are many and complex, it has been exacerbated by the decrease in the availability of fertile land and water caused by the arrival of refugees from the areas in the south of the country, torn by civil war. The resulting higher population density and livestock brought in by nomadic herdsmen, with a consequent increase in demand for resources in an area where the survival of 75 per cent of the population depends on the direct exploitation of the land, combined with climatic conditions that have made the region one of the driest in the world since the 1970s, have fuelled antagonism and armed clashes between nomadic herders and farmers. The availability of scarce land and water, if not the direct cause of the conflict, has created ideal conditions for the resulting explosion and its prolonged duration.[53]

An area where many land deals are taking place is the Tana River Delta in Kenya. Here, the dispossession of land (and water) of local communities is combined with a negative trend in climate. Although the region is unique for the richness and fragility of its ecosystem, including some of Africa's most important freshwater wetlands, the Tana River Delta has been designated as an area of strategic interest for agricultural production and biofuels by the Kenyan authorities. The government in Nairobi has implemented several national projects, one of which is state-owned, involving the intensive cultivation of sugar cane and the refining of the raw material on a total of over 100,000 hectares. It has also granted permits for 20,000 hectares of jatropha to a Canadian company and is negotiating a further 50,000 hectares to be devoted to the cultivation of oilseed by a British firm. In addition, a local company would be assured production of fruit and vegetables on 5,000 hectares, apparently without any written agreement so far but it has already cleared 250 hectares and started planting maize. This perverse tangle of domestic land-grabbing, national and international investments in energy crops and looming land transactions is affecting an area at risk, not only in environmental terms, where the lack of infrastructure and basic services has led to marginalisation, banditry and subsistence levels of existence. The ethnic groups that inhabit the region, Pokomo fishermen and Orma and Wardey pastoralists, do not have titles to land ownership. Back in 2000, following a controversial programme of land allocation, with changes in the existing rudimentary irrigation systems in the region and a prolonged period of drought, communities entered into a low-intensity conflict for access to natural resources. This situation could be aggravated further by any land deals, which would require an eviction programme affecting thousands of people in an area where the effect of climate change is causing severe soil erosion, high rates of evaporation and water salinisation.[54]

International land transactions could also be a source of insecurity in terms of geopolitical balance. From this point of view the case of the band of land covering Ukraine, southern Russia and Kazakhstan is particularly interesting. Located geographically between the western economic block and that of the emerging powers in Asia, it is an area with an estimated 20–40 million hectares of land available, the average price of which is rather low. It is also a region that is rich in energy resources and the oil companies themselves were already beginning to invest in land back in 1999–2000. In the period 2006–2009 the area of cultivated land traded totalled 3.5 million hectares.[55] Kazakhstan is attracting the interest of the Gulf States and its neighbour China whilst in Ukraine most purchases are made using the capital of private companies, especially European investment funds and Libyan sovereign wealth funds. Russia, on the other hand, plays across the board, announcing the sale of portions of land in the north whilst undertaking acquisitions in its two neighbours to the south through companies and investment funds. For its part, in 2007 Moscow launched the idea of a 'grain OPEC', a coordination of the grain markets to be implemented together with the two former Soviet republics, showing an awareness that food and agricultural land resources are crucial for development, perhaps even more than oil. A transnational agricultural power with control of strategic foodstuffs is not yet on the horizon. The project has entered deadlock largely as a result of the non-adherence of Ukraine, the former breadbasket of the USSR which is now split between the European dream and Russian influence, and has been under strong commercial pressure for some time as far as its national arable land is concerned. In

2009 the government in Kiev passed a moratorium on transactions, while new legislation is before Parliament that would ban the purchase of land by foreigners, but not its rent.[56] The phenomenon of large-scale land acquisitions has also worried neighbouring Romania, which in 2014 will eliminate the restrictions on acquisition of land by foreigners in agreement with the EU. The land deals, with their set of unknowns, are knocking on Europe's door.[57]

Best practice

So far, large-scale land acquisitions seem to have brought few benefits, especially in those countries where the concept of land ownership is weak, and where the authorities at national and local levels are not up to handling transactions of great complexity and potentially enormous consequences in a sufficiently transparent fashion.

This does not mean that there is no experience of foreign direct investment in agriculture in which the exploitation of land by an enterprise in a foreign country is part of a sustainable business model for local people. One example is the Mali Biocarburant SA (MBSA), the European biodiesel producer that operates in Mali and Burkina Faso. MBSA employs the intercropping model using jatropha, an inedible tree with a high oil content (which can therefore be used to produce biofuels) interplanted with the food crops grown by more than 8,000 local farmers on over 7,000 hectares in the two countries. The company, funded in part by the Dutch government, has not acquired the land directly but exploits it for the production of biofuels for the local market with the involvement of the farmers who own 20 per cent of the company. They cultivate the jatropha to complement their food crops, collecting the seeds and selling them to MBSA, which uses them to manufacture biodiesel. The residues are then used in the manufacture of fertiliser and soap. The production satisfies the demand for low cost fuel, particularly needed in Mali, where the government has launched a national jatropha programme which limits its support only to those initiatives oriented to the domestic market.

The example is salutary if one thinks of the cause and effect relationship that is often reported as an outcome of policies incentivising biofuels – the exponential increase in land deals. As mentioned above, a factor that fuels interest in energy crops is the possibility for investors to get a return from the sale of carbon credits, through the systems of mandatory reductions in CO_2 emissions, such as the Clean Development Mechanism (CDM) and the European Emissions Trading System (ETS). MBSA is one of the first companies in Africa to benefit from carbon credits, but reinvests resources in two foundations that aim chiefly to improve the irrigation infrastructure, the purchase of fertilisers, the training of farmers on farming techniques, not only for jatropha, but also of basic foods. Hugo Verkuijl, the CEO of MBSA, said:

> Thanks to the foundation we have been able to introduce a new drought-resistant variety of maize to our producers and crop yields have increased from 0.5 tons to 2.5 tons per hectare in a single year. Food production is not the main objective of the MBSA, but it becomes such because without food security our farmers would not grow jatropha.[58]

The use of the plant as a complementary crop also ensures that areas of low productivity for the production of food in rural areas will not be planted up, a condition that provides a complement to farmers' incomes and integrates rural development and the more general growth of private initiative in the local economy. 'Ours is a labour-intensive production,' continues Verkuijl, 'that we want to export throughout West Africa, especially using the technique of the foundations. The authorities in Senegal have already shown interest in our model, because it is economically sustainable whilst, at the same time, reinforcing a system of production based on small-scale land ownership.' In short, offering protection against uncontrolled large-scale, acquisition of farmland which spreads more easily when there is incomplete awareness of property rights. 'Apart from the anxiety that land-grabbing itself generates,' concludes Verkuijl, 'it represents a mistaken model of investment because it can operate in the short term but not over long periods.'

Investment and human rights

The theme of 'responsible investment in agriculture' was put forward for discussion for the first time internationally by Japan at the G8 summit in L'Aquila in Italy in 2009. The debate that ensued from the proposal fortunately assumed global dimensions. The hypothesis brought to the table by Tokyo then took shape in the form of the proposal of a 'code of conduct', a set of criteria to comply with on a voluntary basis for the actors involved in the investments in question. While discussions are ongoing for the adoption of the voluntary guidelines for responsible governance of land ownership in the Committee on World Food Security of the United Nations,[59] the most advanced initiative in this direction is a joint document of the World Bank, FAO, IFAD (International Fund for Agricultural Development) and UNCTAD (United Nations Conference on Trade and Development), which identifies seven principles to be adopted to guide investments in such a way that they respect rights, livelihoods and resources.[60]

This is an approach that has not failed to arouse criticism. In the principles for responsible investment the NGOs see a way of not stopping, but 'legitimising the land-grab' and propose strict international regulation instead. The UN Special Rapporteur on the Right to Food, De Schutter, puts forward eleven principles to be adopted in land policies so that the issue is addressed from the perspective of rights rather than investment. Similarly, some scholars dispute the narrative 'of making a virtue of a necessity', claiming the need to redefine policies, starting out from the land rights of people living in rural areas of the 'global south'.[61]

The two points of view both find room in the legislative procedure underway in the European Union with regard to the new 'strategic framework to help developing countries to tackle the problems of food security' which was the subject of a Commission Communication dated 31 March 2010. If the Executive aims to 'encourage the development of international principles for responsible investments in agricultural land', the Parliament, with the report by Gabriele Zimmer, seeks to 'develop and adopt a strict regulation at national and international level on acquiring land'.[62]

These are positions that reflect broad participation, but also an insidious clash between the technical and political levels, between civil society and the institutions, which could block or delay the necessary answers to all the questions that arise from the issue, ranging from land ownership governance and the quality of investments for development through to the protection of the rights and the functioning of agricultural markets. In this 'dialogue between deaf people',[63] the International Land Coalition acts as a clearing house, which includes 116 organisations including research institutions, local NGOs and global intergovernmental institutions. Its aim is to promote equitable and secure access to land through dialogue and the sharing of knowledge and to establish a monitoring centre on large-scale land acquisitions and their impact in environmental, social and economic terms.

Interview with Michael Taylor (International Land Coalition programme manager)[64]

What is the rush for land? Is it just land-grabbing?

Land-grabbing is only part of the phenomenon we are seeing, that is to say a dramatic increase in land-based investments. But not all land-grabbing is linked to investments such as in the case of local land-grabbing, sometimes perpetrated even within the same family. Then there are investments in which the acquisition is carried out unlawfully. In these cases you can also talk about land-grabbing. But the new thing compared to the past is that we are faced with an increase in agreements that are entirely legitimate in terms of their legal and formal aspect.

But in many of these cases the rights of local people are not taken into account, especially traditional rights to land. What do you call this?
Land-grabbing. But it's not the investor who is doing it, rather it is the local government giving a land concession while knowing that there are people who have been using it to provide their livelihood for generations. Land-grabbing is only part of the commercial pressure on land that is not only related to the agricultural production sphere, but involves sectors such as tourism, mining, the exploitation of forests and so forth ... we can say that there is a very broad front of interests that is leading to an increased demand for land with targets that lie mainly in the southern hemisphere. This increases the risk of land-grabbing, but lends hope for the emergence of different patterns of investment in agriculture, which are so desperately needed.

Initially the phenomenon was studied mainly in relation to acquisitions on behalf of states or by sovereign wealth funds. Recently, attention has shifted to private investors. But weren't they there before?
Of course. A few years ago acquisitions on behalf of emerging economies such as the Gulf States, South Korea and China made the news, new players that were going to upset the balance in a setting in which, historically at least, European and North American companies had made the running. And this is good, because it has made public something that always existed at the private level, although on a much smaller scale. Of course it's a concern that some states, to secure their food supply, should rent or buy land for food production in countries that have been struggling

for years with malnutrition. But we must not forget that most of the acquisitions are by private bodies.

What are the consequences of this?

Many private bodies, especially the hedge funds, like to repeat that responsible investments can be made and can earn a lot in some southern countries of the world because 'the land is cheap'. The problem is that part of their strategy is actually based on there being an appreciation in land, which I think is dangerous for the local economy. If it is true that there is a large margin between the potential value of the land and its market value, to whom should the income derived from this margin go? If it is land that, by customary rights, belongs to the local population, is it right for a company to turn up and appropriate that value in its entirety? One of the recurring critical aspects in agricultural investment that has emerged in recent years is precisely that of the transfer of property rights. But there are other possible models that work and are based on cooperation with local communities.

Three years after the Daewoo-Madgascar affair, what have we learnt about large-scale land acquisitions?

That it is a global phenomenon but that it is only a symptom. It expresses the plight of democracy in most countries where land deals create conflicts with local communities. Places where, in the process of decision-making, there is insufficient scope for people's participation. Some governments are only theoretically representative of the interests of the community, more often they essentially operate in their own. Things are even worse in terms of the governance of land ownership, which in many places simply does not exist in spite of recent land reforms. As we have said, the most controversial investments involve land that is legally owned by the state, but where, by tradition, there are those who have cultivated it for generations. In this case, this is a right that the national and international legal framework does not recognise. Another aspect is the increasingly evident commoditisation of land. We are agreed that it represents a productive asset but in my opinion it cannot be considered a commodity like any other and should not be traded on the market without protection, because it also represents the individual dignity, culture and livelihood, not only locally, but for billions of people. According to IFAD, 450 million small farmers, many of which produce on land for which they do not hold title, feed some two billion people worldwide. Which brings us to yet another critical point in large-scale land acquisitions. They are supported by the belief that small farmers are useless, a hindrance for intensive industrial production which represents the only possible response to the challenge of future food supply. On the other hand, until the 1970s, as far as intergovernmental institutions were concerned, this was the only way forward for agricultural development.

A recipe that many governments of the south are adopting today, just as intergovernmental organisations appear to have abandoned the model of the large company as a vector of development ...
Small landowners, for cultural reasons, are considered by many governments that offer land as backward-looking. It is an ideology that partially reflects the economic interests of elites.

What is the current state of the debate with regard to the steps to be taken to address the problem?
There is the need for investments to improve agricultural productivity, but these must be of the right type. What is increasingly emerging is an awareness that those based on the transfer of ownership or long-term leases may not work. We must work together with the people holding those rights, whether or not they are formalised.

At the moment the debate is between those who would like a code of self-regulation and those who would like to see certain rules in order to bring about a ban on 'land-grabbing'. What do you think?

I do not think it is feasible to establish a comprehensive legal framework with rules that are the same for everyone on this issue. Each country has its own rules on land ownership and it is far from certain that they would be willing to change these in the short term. The first phase should include a set of principles of good practice and reference. Those produced by the FAO, IFAD and the World Bank are a starting point, but have been written without consulting any representatives of civil society in the countries most affected by the problem They are imposed from above. Even if the contents are correct, there has been no genuine process to arrive at a consensus and therefore these principles, in all probability, will remain on paper. A process that is more open to contributions from the bottom up is that of the guidelines of the United Nations' Committee on World Food Security, on which many people are pinning their hopes. The consensus on the rules to be adopted should be built with the participation of civil society. The balance of power necessary to get rid of land-grabbing comes from providing a voice to local organisations so they can independently influence the choices of their countries' governments with regard to the rights of land ownership. Let's not beat about the bush: this is going to take time.

PART III

A global policy for food security

6

STARTING FROM G20

In the preceding pages we sought to clarify the origin and implications of the food rush and the era which we termed that of new scarcity.

Consumption grows and changes, while the availability of natural resources is increasingly constrained. Focusing on the case of land, we have seen how its scarcity even pushes states to consider the purchase of land beyond their borders with controversial implications, especially for developing countries, which end up giving up pieces of their most fertile land at bargain prices. The new restrictions on the use of basic resources like water and the need to make farming more environmentally sustainable paint a picture that is not encouraging in terms of food production, the colours of which are rendered even more dismal by the predicted effects of climate change which is destined to raise many more vulnerabilities in the food supply system.

The changing scenario, the full extent of which is only now being assessed, is putting a strain on the responsiveness of our economic and political systems. In particular the responses by individual states do not seem to be up to the job, and even the international community has found itself unprepared on several occasions when the lack of coordinated action at a global level has triggered reactions that have merely served to reinforce the volatility and further distort the transmission of signals in the market. We are talking about a market that, we should remember, is in itself rather unusual. Low trading volumes, fragmented supply structure and high uncertainty are all factors that limit farmers' improvements in technical and organisational capacity across most of the world, obviously with varying intensity depending upon the level of development.

The events of 2007/2008 and those of 2010/2011 are unfortunately more than an alarm-bell. The most recent spike in food prices alone has shifted almost 45 million people below the poverty line, synonymous with leading an existence on less than 1.25 USD a day, which is already experienced by more than 1.2 billion people around the world. The tensions between supply and demand suggest this is a structural imbalance that will grow in the years to come. The failure to meet the first of the millennium development goals, a reduction in hunger in the world, ought to restore a sense of urgency and determination in order to conceive and design renewed policies at the international level, to build a new overview of global food security. Given the scale of the problem, this need cannot be relegated solely to its traditional place on the agenda on just how to support agricultural development and food self-sufficiency in late-developing areas, but should be tackled with decisions that necessarily involve agriculture worldwide. It is a problem that affects us all and is not just a question of how to help 'others'.

After decades in the wilderness it is no coincidence that the issue of food security is once again part of the lexicon of American and European policy-makers, just at the moment when the USA and EU are preparing to reform their own agricultural policies, the longest established and also the most criticised.

It is also no coincidence that the G8 Summit in L'Aquila in Italy in 2009 and the G20 under the French presidency in 2011 chose to raise the issue of food security as a global problem on the international political agenda which, with varying intensity and urgency, will affect all of us, at least in the medium to long term. The most concrete result of the summit of agricultural ministers in June 2011 was approval of an Action Plan to cope with price volatility and to support agriculture.[1] The document identifies the main areas for intervention and provides information of a general nature, much of which is aimed at financial markets. The plan was by no means universally acclaimed. The French Minister Le Maire called it a success, as did all the leading players. According to Tom Vilsack, US Secretary of State for Agriculture, the summit was a historic achievement. To balance this enthusiasm, there was the dissatisfaction of many international NGOs calling for more concrete commitments, and the World Bank President, Robert Zoellick, for whom the G20 agricultural summit had had a 'modest' outcome. Most scholars and analysts were also cautious, finding the document too unbalanced on the theme of the 'financialisation' of agricultural markets.

When all is said and done, we believe that the Action Plan can be considered a first fundamental element onto which to graft a plan for coordinated international commitments. A first step that has had the effect of bringing the issue to the attention of world public opinion, helping to raise awareness that the scale of the problem of the stability of food supplies is different from that to which we were accustomed in the past.

The challenge is being played out on at least two levels and two timescales. On the one hand, we have to work on research and technological/organisational transfer and, on the other, on a review of national and supranational policies that govern trade and food security. A massive dose of innovation is the only way for what has been called the prospect of 'sustainable intensification' – the ability to produce more without consuming new land while polluting less. As far as policies are concerned, they can do much to ensure greater efficiency and stability in the markets, to protect farmers from heightened uncertainty surrounding their activities and to prevent widespread instability falling on the weakest sectors of the population resulting in social tensions that will prove difficult to control. The policies have a role to play in the first area of intervention under consideration in promoting the spread of technical developments, especially in parts of the world where levels of efficiency are lower and the food emergency more acute.

There are also two timeframes within which to design new governance tools regarding the food supply, short and medium–long term. In the near future we must find solutions to deal adequately with the problem of market volatility and the situations of food vulnerability associated with it and its impact on the agricultural production fabric. In the long run, we need policies able to provide lasting and generalised food sustainability.

Contributions

Policy-makers taking responsibility

Martin Schulz – President of the European Parliament

The world is undergoing extraordinary changes in demographic, climatic, ecological and economic terms. This is a period in which the imbalance between the exploitation of resources and their capacity for regeneration has assumed an increasingly important position on the international political agenda and has opened a debate that affects not only the short to medium term, but the very conditions by which we are building and ensuring the well-being of future generations. The effects of this imbalance are, however, already making themselves felt and are often impacting heavily on the economic and social conditions of the inhabitants of the planet. We have entered a historical phase that requires us to urgently address the issue of the sustainability of growth processes, both demographic and economic. We find ourselves at a moment in time in which we need to rethink, at least partially, the trajectories of development, in order to render them more balanced and consistent with the challenges facing our society in the years to come.

The rise in food prices experienced in 2007/2008, in late 2010 and earlier this year have aggravated the situation in the poorest areas of the world, where the bulk of incomes are used in food purchases. In this situation everyone must take responsibility, starting in Europe which today is not only the largest food producer in the world, but also the largest importer of foodstuffs and exporter of food. This position should push us into playing a leading role in the challenge of achieving stability in the food supply. It is a complex issue, which should enable us to deal with the intolerable emergency created by a billion hungry people in the world and look forward, in the long run, to allowing us to meet the growing demand for food while minimising the environmental impact.

I use the word 'we' in referring to the European Union, but also the political class in general, because their role will be crucial in managing this phase. When it comes to agriculture and food production, we cannot ignore the specific nature of the sector, which is of great public interest and requires a greater degree of involvement on the part of policy-makers. This, unfortunately, does not always happen because sensitivities at the international level vary considerably, even on important issues such as climate change, and is the reason why, in recent years, politics has found itself wrong-footed on some occasions by the way in which some markets have evolved and has found itself unable to regulate them to prevent improper conduct.

In certain markets, such as agricultural ones, in which the manufacturing base is often organisationally and structurally weak, the role of the policy maker takes on a greater burden of responsibility. Indeed, this is why public intervention has always been a feature of the history of agriculture throughout the world. Politics has the task of intervening on the supply side, promoting research and development, but especially in the transfer of innovation, advancing the goal of 'producing more and polluting less'.

Politicians also have the responsibility to create the conditions to ensure that agricultural enterprises are economically viable and sustainable, providing a source of income for the farmer to curb the phenomena of environmental and social degradation triggered by the abandonment of this activity. This sort of support must now establish a suitable dimension and efficacy, especially in areas where access to food is now a matter of survival. But even the more developed, or so-called emerging markets, can and should make an important contribution to the growth in supplies over the long term.

But now politics finds itself with the task of intervening in the markets and quickly overcoming the failure of the Doha negotiations and restoring the credibility of the WTO. Of course, the scenario today requires us to revise somewhat the idea of how markets function, which had become established until a few years ago, and there is an urgent need to promote their greater transparency, as was outlined by the G20 meeting in Paris. Financial speculation, such as futures in food commodities, has played a role in the recent price spikes. Investors should not be allowed to play roulette, gambling on food prices, and the EU should take steps to prevent this. But the question of greater international coordination of measures also needs to be addressed and we need to promote the formation of adequate reserves of strategic raw materials and prevent unilateral actions (such as export bans) aimed at promoting food security within particular areas that end up destabilising the global food supply and increasing the size of those populations at risk from hunger and malnutrition.

More research and greater courage to change

Joseph Daul – Chairman of the European People's Party Group in the European Parliament

Over the centuries we have seen the role that rising food prices can play as a trigger for riots and unrest, and those who think that our society can now consider itself immune from this kind of tension are committing a very serious mistake. Experts have been warning us for a few years now with regard to the risk that demand for food might outstrip supply in the near future, but are we really listening?

As we know, estimates tell us that the world population, which has now reached 6.8 billion, will exceed 9 billion people by 2050. The volume of agricultural production on the other hand is contracting in a trend that will become increasingly evident forty years from now. According to the FAO, agricultural production rose by 2.3 per cent a year from the 1960s onwards, but this figure will decline to 1.5 per cent by 2030, and 0.9 per cent between 2030 and 2050. Yet, to feed the population, we need to produce 70 per cent more food globally, a figure that reaches 100 per cent in developing countries.

This is a time bomb in humanitarian, political and ecological terms, a threat to the stability of the planet itself unless we begin to defuse it immediately. Personally, I think that we will succeed eventually, but this challenge requires strong political leadership at global level and the willingness to accept that things must change as well as a realisation that new approaches are needed. The surge in food prices we witnessed in 2007/2008 had already brought food security issues to the top of the political agenda and the new price records reached this year to mark events demonstrate the need for urgent action.

In June 2011 we saw the commitment to act at a global level put to the test when, for the first time, the agricultural ministers of the G20 nations met up and drew up a plan to halt the price rises. Now it is up to the leaders of the G20 to implement the recommendations contained in that plan. There is an urgent need for greater transparency in financial markets that trade in agricultural commodities as well as a serious commitment to addressing the abuse and manipulation in this sector as it is causing suffering to the poorest on the planet. Early warning systems are essential, so that we do not find ourselves unprepared in the event of a sudden increase in prices, as has happened in the past, and we need to think very carefully about appropriate ways to ensure that food stocks are available where and when they are needed.

Fundamentally, we must produce more food. This is the biggest challenge. We will have to achieve this under conditions that are more difficult than ever, since the cost of inputs such as fertilisers and energy is continuing to grow and climatic conditions seem likely to become less favourable. Rich states must help those in developing countries to promote agriculture, for example by reducing the difficulties small farmers have in accessing the loans and micro-credits they need to be able to invest in seeds, fertilisers and efficient irrigation systems. Rules on the ownership of land need to be introduced where they are lacking, to ensure that the sale or lease of agricultural land to foreigners does not mean a further undermining of the stability of food supply for local communities – a concern that my friend Paolo De Castro rightly accords great importance.

The European Union is providing its contribution to global food security. We are leading the world in *per capita* contributions to development and recently created tools worth a billion euros to be able to respond quickly to increases in food prices in developing countries. That said, we must do more. We must invest more in research in order to increase productivity and also make the results of this research available to the poorest countries. The majority of economic aid aimed at them needs to be directed towards agriculture because it is through efficient agricultural markets that wider comprehensive and lasting economic development can be ensured.

European agricultural production also has to increase to keep pace with the rise in demand. The European Union is the largest net importer of agricultural commodities and, something always neglected by critics of the Common Agricultural Policy (CAP), the largest share of our imports comes from developing countries, from whom we import more food than Canada, the USA, Japan, Australia and New Zealand combined. In all, 71 per cent of our total agricultural imports come from developing countries as a result of initiatives such as Everything But Arms (EBA), a generalised system of preferences and economic partnership agreements.

These are all factors that need to be taken into consideration during the drawing up of the new CAP to come into force after 2013, which will take shape in a context that was barely imaginable a few years ago. The issues of food security, price volatility and the economic crisis are dominating the media agenda worldwide. Since its inception, the CAP has proven to be a policy that can transform itself to achieve the required results. The new CAP will have to satisfy the legitimate concerns of citizens with regard to its environmental impact, but will also have to continue to ensure a stable food supply of good quality at reasonable prices. These two requirements are not mutually exclusive but, as a result of the European Parliament's role in decision-making, appropriate policy choices have to be made. We cannot afford to increase the bureaucratic burden on farmers, making them think that farming is no longer worth their while. This would be disastrous, as would the accepting of trade agreements which would see the uprooting of agricultural production in some parts of the world to concentrate it in others.

Finally, I would like to return to science and technology. We need to keep an open mind with respect to the possible role that biotechnology can play in ensuring food security, especially if worsening climatic conditions are aggravated by increasing water scarcity. We will need to be able to depend on the best possible results of research to be able to make the correct choices.

An ever greener revolution

More than two-thirds of the world's population living on less than a dollar a day live in rural areas. These are areas where agriculture is the main, if not sole, source of income. Increasing the technical efficiency of farmers in these regions means increasing their incomes, but also the self-sufficiency in food of less developed countries.

With less than two hectares per capita, small farmers account for almost all economic activity and enterprise, supply and demand, production and consumption in these regions. In some cases small-scale agriculture amounts to more than 80 per cent of the food production in some developing regions. These people are therefore major actors in triggering the processes leading to economic growth. As a result these local actors should also be decisive for investments in agriculture in developing countries. The technical and organisational growth of agricultural systems falls by necessity on their shoulders and can best be configured only with the coherence of their initiatives.

To do this we must raise their knowledge and skills level, improve the quality of services and their ability to access them, calibrating the research and transfer of innovation on the specific aspects of the environmental and production conditions in the areas in which they live.

Improving human capital is a prerequisite. Keeping pace with these markets means having to interpret these multiple phenomena and translate this interpretation into management initiatives that tend to become increasingly complex, requiring continuous updating to take into account both technical and organisational innovations. This objective implies a greater incisiveness of gender policies: women make up over half of small farmers around the world and enhancing their role can be of great significance in terms of the stability of food supplies.

Raising the sum of human capital is one condition. Equally important is the contribution they can make to the development of the services and infrastructure starting with irrigation systems. The FAO estimates that about 1.2 billion people live in regions with marked problems of water scarcity and the situation could worsen in the short term, coming to affect another 600 million people by 2025. The advantages gained through the establishment of new and efficient irrigation infrastructure could be immense, particularly in Africa. The productivity differential is approximately 130 per cent between irrigated and non-irrigated areas, and in places where water is not only a problem of distribution but also with regard to the volumes available, updated irrigation techniques can help to achieve major results. The transport and storage infrastructure should also be supported and fostered. Farmers in the poorest areas of the world sell their product at harvest when the price is at its lowest, having to buy it back when the cost is higher, contributing, paradoxically, to a general increase in prices. This is due to a lack of appropriate technology and facilities for handling, storage and processing, which would extend the shelf-life of the product and raise its value in the market place. Finally, there is the need to strengthen the credit system and, more generally, risk management. The poverty in which farm families live represents a barrier to investment that becomes insurmountable in the face of the slightest adversity, such as falls in production, lower prices or increased energy costs. Unfavourable years often affect the continuity of farming, making it impossible to maintain operating costs at normal levels and in such a situation the ability to borrow or have access to tools for income stabilisation[2] is essential in the short term, to ensure that farms are economically viable and are able to respond more consistently to market signals.

Then there is the great theme of innovation transfer. Investment in research needs to be combined with efficient technology transfer services and assistance and with appropriate market incentives for their introduction and use in order that the benefits of innovation are able to reach the greatest number of farmers, ensuring that technical advances can be adapted to local conditions. In this area, as already mentioned, there remains much to be done with there being technologies already considered obsolete in more developed countries that have not even begun to be applied in the agriculture of some of the poorest areas of the world. The gap is still much too high. In North America, maize yields reach ten tons per hectare while the highest in sub-Saharan Africa is two tons, three in South Asia. We have to ensure a rapid rise in productivity in these vast areas, primarily to ensure that there is a local response to reducing hunger. Already proven and available technologies that are more environmentally sustainable need to be made available, especially in areas where water shortages are greatest. There are important examples that already go in this direction such as the one modelled by the initiative of the US Government 'Feed the future: global food security research strategy',[3] which refers to the investments in systems that protect water and soil using nitrogen-fixing plants in South Africa. The systems introduced have led to significant increases in yields for some 300,000 farms, enabling them to improve the soil's ability to retain water and nutrients in arid soils. Obviously, all this requires larger investments and more effective transfer of skills and technologies.

Improvements in research should not only allow us to achieve the important goal of completing the Green Revolution, with the ethical obligation and the economic connections to make it even more 'green'. Experimental activity must broaden its horizons, looking further afield to examine long-term sustainability, to allow agriculture to develop in the most fragile environmental contexts, hence the need to develop techniques that are less polluting but at the same time more productive, able to cope with an expansion in the areas subject to water scarcity. In the field of agricultural science the slogan 'more crop per drop' has become widespread and is used to indicate one of the major agricultural research requirements, especially in the sphere of applied plant genetics.

Encouraging research remains an objective to be pursued with greater intensity in more developed countries as well. The improvements in efficiency that can be obtained on the basis of existing knowledge and progress already achieved enable acceptable results to be achieved more quickly. As far as the financial requirements in this field are concerned, both current economic powers and emerging ones need to drive innovations and their application in agriculture.

Therefore we need to set about reversing the trend of recent years, and public investment in agricultural research should become a priority for countries throughout the world, as well as encouragement of private investment. We have already discussed the peculiarities of the interactions between public and private investment. Here, however, we need to stress that the challenge of food security implies long-term vision and the use of these funds can only be ensured by a massive, well-targeted public intervention.

Today, as in the past, we have to turn to research and continued progress to disprove the theories of Thomas Malthus. The technological frontier has to do more than just walk forward over the coming years, not only in terms of quantity but also in reducing the environmental impact of farming, and along this route the role of policies can prove essential, for example through stimulating private research. The definition of limits on the emissions of pollutants, incentives for the adoption of ecologically sound technologies, as well as a more leading role to be given to the market to reward the production of environmental goods,[4] facilitating their recognition by the consumer, are all factors that can promote private research and direct it towards the output sought by society itself.

We need to approach these issues with no ideological prejudices. We refer in particular to biotechnologies employed in agriculture, which can make an important contribution to increases in productivity and the achievement of food security objectives. This is especially true in the most fragile areas of the world where epidemics, climatic shocks and water scarcity constraints are crucial factors in the development of agriculture and the economy in general. The CGIAR, the Consultative Group on International Agricultural Research,[5] emphasises the role that biotechnology can play and, more generally, the role of plant breeding in increasing yields and farm incomes in the poorest areas of the planet. But the FAO, the World Bank and other reputable international institutions involved in combating poverty say the same. Genetic engineering is an opportunity to expand traditional or 'conventional' – as critics of GMOs like to call them with a touch of linguistic malice – techniques decisively, and developments in the laboratory in this field have produced cultures resistant to climatic and water stress and some of the most common plant diseases, while allowing reductions in the use of pesticides and other pollutants.

We are not saying that everything is rosy. But it would be short-sighted to deny the potential of GMOs. In general, their use raises concerns not only about health and environment, but also about accessibility and the transfer of innovation. A modern policy must not be afraid to tackle these concerns, as unfortunately happens in many countries, but should rather respond to them on a case-by-case basis. For this reason we believe that the role of public research in this field is essential in allowing everyone access to technical progress and improvements in their standards of living and we need to ensure that biotechnologies are at the service of all and are consistent with the needs of society. Equally important is the role to be played by access to information. We cannot reduce decades of scientific progress to a biased and radicalised 'yes' or 'no' public debate relying heavily on fears that could be justified at the start given the uncertainties that inevitably accompany the exploration of new areas of applied research. Science has made great strides and we now know for sure that the risks of some genetically modified crops are equal to those of conventional ones and the only aspects that are not 'conventional' are the benefits with regard to productivity and, in many cases, the environment.

The rules on trade

In a recent article published in the *American Journal of Agricultural Economics*,[6] two leading scholars argue that in

2008 the trade policy initiatives aimed at reducing the impacts of price increases were, in fact, a major stimulus to the increases that immediately followed. In particular, the adoption by many states of measures to restrict or prohibit exports and the reactions implemented by net importing countries that sought to stimulate and facilitate imports accounted for 45 per cent of the price increase for rice and about 30 per cent of that observed for other cereals.

The responsibilities of politics in this respect are great, and remain so: the same behaviour was recently adopted even if it went on to produce the opposite effect to that desired. Perhaps interventions of this kind were inevitable but, as stressed by a recent work of the International Centre for Trade and Sustainable Development,[7] governments and international institutions were not prepared to cope with price movements of such magnitude. The world was taken by surprise and quickly catapulted into a new and less reassuring scenario. The absence of appropriate intervention tools in the supranational policy solicited 'homemade' and totally uncoordinated responses to defend short-term national interests.

The challenge now is to rethink the rules and the role of international trade policies, armed with the knowledge that we are not dealing with a transient phenomenon but a structural one which introduces us to a historical phase in which we have to deal with higher prices and more volatile markets. If the document released by the G20 can be viewed as a step forward in defining an agenda for global food security, it should be noted that, to date, there have been no innovations in the international institutional framework that might permit us to state that a possible new price crisis could actually be managed any differently from those in the past. In the short term it is legitimate to expect an agreement not to employ any of the prohibitions and restrictions on the food purchased for the World Food Programme. Yes, you read that right. Even humanitarian aid has been and is subject to trade restrictions adopted by exporting countries during times of crisis.

All told, it still seems too little is being done.

We need to act urgently, starting from the reorganisation of trade policies, initiating the necessary reform of the rules on international trade practices relating to restrictions on exports. We have to redesign the framework within which to re-bundle and contain these measures, combining them with other means to contain domestic inflation, such as transfers of food for social purposes only in exceptional cases and as an alternative option, a question to which we will return shortly. We need to define the cases and the duration of any restrictive measures more clearly, placing them within a system of administration of emergencies that is integrated at an international level.

This, however, is only part of a larger project which covers the re-launch of the mission of the World Trade Organization (WTO), that is to open markets and make them more accessible and efficient. On this front as well there are no encouraging results to report, indeed, far from it. International trade negotiations have recorded virtually no progress since the beginning of the 1990s. The current round seems destined to represent yet another failure and this paralysis contrasts with the speed of change taking place, to the extent of calling the credibility of the WTO itself into question. The obstructive techniques employed by the blocks and the commercial interests that divide countries and developing or emerging nations must be overcome in the light of new emergencies.

Talks need to be re-launched on a new basis. The G20 commitment should be a political stimulus for a new undertaking to negotiate. The first objective must be to accelerate the process of the liberalisation of agricultural trade, only by reducing the barriers that currently restrict trade flows will markets increase in size, avoiding the current situation whereby small changes in the quantities of foodstuffs available correspond to large swings in prices. By precisely embarking on this road one can also obtain an increase in efficiency in the transfer of farm commodities from areas with a surplus to those with a deficit. The first and most important 'medicine' is, therefore, the integration of agricultural markets. But on its own this is not enough and the cure should be complemented with other instruments and especially with the idea of making the institutional framework more flexible and able to respond promptly and effectively to emergencies that occur again in the near future.

More transparency, more efficiency

A primary and essential pillar of any system of crisis management market is information. The lack of reliable data on production, demand, stocks and volumes available for export has fuelled uncertainty and anxiety at times when prices are rising. This is one of the basic principles of the functioning of markets: their efficiency increases with the spread and completeness of information. With the phenomena we are witnessing, the absence of detailed information on key market variables has contributed to reducing their efficiency, accentuating price movements. In future, the availability of accurate and up-to-date information represents one of the most important tools in tackling this volatility.

The reasons that have so far prevented the creation of adequate information systems are manifold. The first is that the data, such as that relating to national stocks, may prove very sensitive. In addition, with regard to stocks for example, changes in agricultural policies, particularly those in more developed countries, are leading to reductions in publicly held stocks, a process which is proceeding at a range of different speeds and intensities that are difficult to follow precisely. As far as private stocks are concerned, the collection of information is hampered by the highly fragmented nature of the production system, while data collection on other market variables is equally complex. To overcome these difficulties, an effort needs to be made to share 'sensitive' information, including that concerning the financial markets in agricultural commodities through the construction of accessible and updated systems for the management and interpretation of phenomena on which a large number of variables have an effect.

This should help all players to move in the right direction, from policy-makers to farmers, and should enable the establishment of the best domestic and international responses to any emergencies.

The issue of transparency is also essential to address the role played by financial speculation. We have already established how many doubts surround the role it plays in price rises. What is certain is that the transparency of transactions can act as a brake and the definition of more precise rules on the volumes traded can restrain the impact of speculation, including short-term movements. Transparency and more regulation in the financial markets can help them perform their function of hedging the risk to farmers, mirroring as much as possible actual economic systems and thus ensuring that price movements reflect the fundamentals and are not the result of panic and misinformation.

Working hypotheses on reserves, biofuels and help for the poor

The responses of governments should, however, be able to rely on short-term instruments capable of preventing market turmoil and moderating its impacts on society.

On this front, there are different working hypotheses on the table and many of them rely on the ability of reserves to stabilise the market. But caution is required and if experience teaches us that the reserves have contributed for over half a century to curb instability, it also tells us that this was another era, which we have defined as one of abundance, in which the role of stocks was often completely the opposite, namely to support the income of farmers in some countries by keeping prices artificially high. This is an attitude which for years has obscured the signals of the market, helping to compromise its efficiency. Reserves today should be conceived in a totally different fashion. In the first place they should be sized according to the actual requirements of the international market, but this is an almost impossible goal to achieve. This would mean being able to determine a desirable price range for each product, which means continually updating estimates on equilibrium prices. In addition, the reserves would need to be distributed between different countries or geographical areas, possibly depending upon specific area needs, which vary from year to year along with the harvests. This would be without taking difficulties into account, perhaps related to repeated crop failure. In such cases it would be difficult to replenish stocks without affecting the volumes traded on markets. In addition, the costs associated with the storage of agricultural raw materials, their transportation and their administration in some cases can prove prohibitive.

One conceivable solution could be found in the creation of an international system of emergency supplies, organised at a macro-area level. This would mean defining a minimum level of global stocks, to be distributed according to the intensity of regional shortfalls, thereby making the operation not too distorting for markets near where the food is required and relatively inexpensive. Reaching this threshold of reserves could be helped, if properly encouraged, by recourse to private storage, which, in the past, has proved to be less costly than state-run stores and more receptive to market trends.

Another area of intervention is that of incentives for biofuels, a sector on which there are many converging interests, including social interests, and very sensitive ones too, as testified by the extreme caution with which the G20 action plan addressed the issue, referring to the need for further study in this field. Biofuel policies involve themes such as energy security, combating climate change and, of course, food supply, and we must also be cautious when taking clear-cut positions. However, we believe that we can agree on a principle of common sense, ensuring that incentives for biofuels are more efficient, rewarding the energy yield of the raw materials in question and introducing mechanisms to reduce and/or set limits on their subsidy to be activated during emergencies. If the justification for incentives for the production of biofuels is indeed to compensate farmers and other actors in the supply chain for the production of public goods (energy security, climate change), then it is quite reasonable to reduce this compensation when it risks threatening another public good that is equally important, in this case the possibility of access to food. It is even more so if one thinks that the growth in biofuel production, driven by incentives, risks reducing the food supply from the agricultural sector in a structural fashion in the medium term. If the choice lies between feeding a person or filling the fuel tank of a car, there should be few doubts about which steps to take.

Other hypotheses have found ample space in the debate that accompanied the work of the G20, such as the idea of developing social transfer mechanisms coordinated at the international level, which would be able to mitigate the impacts of food inflation on the most vulnerable in the population.

Already today, many agricultural policies, including those of more developed countries, provide aid to the needy which is supplied according to the cost of staple foods. There are a range of different mechanisms, which very often and not always consistently, have the twin objectives of social transfer and support to farmers.[8] Very often they also have the advantage of representing key buffers in emergency situations. The system of subsidies to consumers who are most exposed to food vulnerability must now be made more effective and coordinated at the international level, as suggested by a recent work by Tim Josling.[9] According to the professor at Stanford University, and we are fully in agreement, it is time to investigate the complementary aspects of domestic policies and international aid in managing social transfers, and identifying target groups and their specific needs. In other words we need to start building a comprehensive and coordinated approach to food subsidies.

The basic idea is to work quickly to draw up a system of multilateral rules for the governance of food emergencies in the immediate future and to make markets more efficient in the long term. This presupposes a strong awareness of the implications that a global structural imbalance between demand and supply of food can produce as well as overcoming traditional trading blocs that until now have hindered the development of international agricultural markets. It also means revisiting some policy choices, such as the incentives provided for biofuels, and establishing international coordination mechanisms that have authority and flexibility by way of prevention or that can intervene rapidly in situations of serious tension.

Managing the risk

Research and trade policies alone are not enough. An intervention truly capable of supporting growth in agricultural supplies over time needs to build on the active contribution of farmers themselves by putting them in a position, in all parts of the world, to carry out their activities in an environmentally sustainable and economically viable fashion. Today, the primary sector is called upon to assume greater responsibilities with respect to the issues of climate change, water conservation and the maintenance of biodiversity. But the same sector has to do just as much with food production. We have seen that research can provide considerable help in reconciling these two objectives, but it is not sufficient. The characteristics of the agricultural fabric, its exposure to risks related to natural events and the instability of the market still require a significant intervention on the part of the 'public hand'. The alternative is the risk of a real disaster, a reduction in the productive potential of the planet and its concentration in the most fertile parts of the world where it is economically cheaper to produce.

This goes for all areas of the world. Not necessarily must the defence of the agricultural potential lead to protectionism and the distortion of free trade, as happened in the past. We must renew the intervention strategies according to certain internationally agreed guidelines consistent with proper and fair functioning of markets. The incentives must drive the adoption of practices and technologies to increase yields and have less impact, compensating farmers for the environmental benefits they produce. In other words, support policies should be directed towards the stabilisation of incomes rather than the market, using intelligent, flexible measures. Among them a special role needs to be played by the tools of risk management. On the one hand, active behaviour in the industry needs to be promoted, such as investing in protection mechanisms right from the start. On the other, access to the opportunities offered by traditional devices for risk transfer to third parties, such as insurance, should also play their part. In parallel with this, institutional intervention should not give up encouraging more innovative options for a large sector of world agriculture, especially in less developed contexts, such as financial instruments for risk management. Incentives should be preceded and accompanied by the transfer of knowledge and business skills enabling small business systems to gain access to the tools that are still too complex.

In general, an urgent effort is required to update the policy instruments available at present, so that in every part of the world a stronger relationship is created between agriculture and society's expectation. The potential of agriculture, still an extremely widely distributed activity even in developed countries, needs protection, especially as it is likely to contract rapidly under the impetus of reforms established in what was another era. It is not such a long time since the last reform of the Common Agricultural Policy (which took place in 2003) but since then the world has changed completely. The really big test for policy-makers today is to bring up to date the reasons for, and manner of, its intervention in agriculture quickly and far-sightedly, in the knowledge that what we decide today will have important consequences for our future and that of the generations to come.

Between protection and liberalisation

We believe that an important opportunity to test this awareness is the imminent reform of the two most important agricultural policies in the world: those of the European Union and the

USA, against whom for years emerging and developing countries have been pointing a finger.

Excessive protectionism and distorting practices in use in Europe and the United States to support agriculture, especially in the past, were a factor impeding the development of international agricultural trade with complex and damaging consequences, especially for areas with less developed agriculture. Without going into detail we can say that over the last twenty years, huge strides have been made in the direction of creating more equitable agricultural policies worldwide, especially in Europe; a path that has progressively led to the removal of the policy's most distorting elements, and, in 2003, with the reform sought by the then Commissioner for Agriculture Franz Fischler, sanctioned a move that many have described as 'epochal', definitively establishing the principle of support for farmers which must necessarily be associated with pro-active behaviour towards society in terms of environmental protection, animal welfare and food safety.

Today this principle must be maintained and extended beyond Europe. It thus appears necessary to construct a paradigm of assistance that can encompass the issue of food security in a manner that differs from what took place after World War II both in Europe and the USA. At the time, the issue was treated solely on the basis of national interests and was enforced using the weapons of protectionism and isolation of domestic prices from international markets. Today, what is required is global vision, and the two largest agricultural powers in the world cannot ignore this. They need to return to provide great impetus to research and development in agriculture, but they also have to construct a new paradigm of intervention through which to support farmers. This requires flexible schemes, which are adaptable to different market conditions, but that above all can safeguard the continuity of agriculture without impacting on the functioning of markets. This means support being decoupled from production, which may represent a real contribution to the stabilisation of farm incomes, protecting them at least partially from catastrophic climatic events and market turbulence.

The reform of the Common Agricultural Policy is already under discussion and should take place in 2013. At the same time the new framework of US intervention for American agriculture, the Farm Bill, should be approved. The responsibilities resting on the shoulders of these two reforms is great. In the era of food insecurity these two policies can lend a new impetus to the liberalisation of world agricultural trade and represent a possible model for action to promote an agriculture capable of increasing production potential whilst protecting the environment. According to classical political thought this seems something of a paradox: how to liberalise and protect at the same time? As we have said from the beginning we do not like to think that certain challenges are impossible. If, amidst great difficulties, farming is gearing up to loosen the double bond of increased production and a concomitant reduction in environmental quality, then the policies that have historically supported it need to solve a somewhat similar problem: a problem which requires untying the old knot juxtaposing protectionism and liberalisation. In the meantime, we have to allay the traditional fears that prevent greater openness, transparency and integration of agricultural markets as well as give support to farmers through the development of new tools, adapted to a changed landscape.

NOTES

Chapter 1

1 There are hundreds of definitions of food security in the international literature. Here we refer to that supplied by the FAO in 1983, according to which food security is the possibility to ensure that every person has access to the necessary food to satisfy their basic nutritional needs.

2 In October 1973 OPEC, the Organization of Oil Exporting Countries, adopted an oil embargo in response to US support for Israel in the Yom Kippur war. The reduced availability of oil on markets led to a sharp rise in prices of energy commodities which led to inflationary growth at the international level.

3 The consumption increase in India, where 40 per cent of the population are vegetarian, will be less pronounced.

4 Namely *How to Feed the World in 2050*, presented at the FAO world summit on food security, Rome, 16–18 November 2009. Available online at: http://www.fao.org/fileadmin/templates/wsfs/docs/expert_paper/How_to_Feed_the_World_in_2050.pdf.

5 See, amongst others, the report commissioned by the British Government entitled *The Future of Food and Farming: Challenges and Choices for Global Sustainability* and *Food Security Assessment, 2010–20*, produced by the US Department for Agriculture.

6 As stated in 2007 by the UN Special Rapporteur on the right to food, J. Ziegler. G. Ferret, *Biofuels 'Crime Against Humanity'*, BBC News, 27 October 2007. Available online at: http://news.bbc.co.uk/2/hi/7065061.stm.

7 Namely *Foresight – The Future of Food and Farming*, The Government Office for Science, London 2011. See also Note 5. Available online at: http://webarchive.nationalarchives.gov.uk/+/http://www.bis.gov.uk/foresight/our-work/projects/current-projects/global-food-and-farming-futures/reports-and-publications

Chapter 2

1 See P. De Castro, *L'agricoltura europea e le nuove sfide globali*, Donzelli, Rome 2010.
2 Particular reference should be made to the Nitrates Directive, with which the EU has set strict limits to substances released into the soil by agriculture.
3 See http://www.aatf-africa.org/wema.
4 The Intergovernmental Panel on Climate Change was established in 1988 in the framework of the United Nations with the aim of studying global warming and its impact.
5 *Navigating the Numbers: Greenhouse Gas Data and International Climate Policy*, World Resources Institute (2005). Available online at: http://pdf.wri.org/navigating_numbers.pdf.
6 The Kyoto protocol is the international treaty concerning global warming signed in the Japanese city of the same name in December 1997 by more than 160 countries. Under the treaty, which came into force in February 2005, after ratification on the part of Russia, the industrialised countries undertook to reduce their greenhouse gas emissions, bringing them, by 2012, to an average reduction of 5.2 per cent compared with 1990 emissions.
7 According to what is proposed by the *Joint Declaration against Food Waste*, adopted on 20 October 2010 by representatives of the EU Parliament, academics and delegates of organisations of cooperation for development (see http://www.lastminutemarket.it/media_news/dichiarazione). In January 2012 the European Parliament approved a resolution entitled 'How to avoid food wastage: strategies for a more efficient food chain in the EU'. See Dongo D., *Il cibo viene sprecato in ogni fase della produzione e del consumo: in Europa fino a 115 kg a testa all'anno. Il Parlamento UE cerca soluzioni*, Il Fatto alimentare, 1 August 2011. Available online at: http://www.ilfattoalimentare.it/spreco-alimentare-il-parlamento-europeo-propone-una-strategia-per-limitarlo.html.
8 A term used by A. Segrè, *Elogio dello –spr+eco, formule per una società sufficiente*, Emi, Bologna 2008.

9 See the sentence of 13 April 2011 on case T-576/08.
10 This is exemplified by *Feeding America* (www.feedingamerica.org).
11 See www.lastminutemarket.it.
12 See www.approvedfood.co.uk.

Chapter 3

1 The cost of the policy would actually be too high because the insurance company would have to tie it to the possibility of compensating too high a number of customers (farmers) in the same year in which, for example, a drought or hailstorm occurred.
2 OECD/FAO (2011), *OECD-FAO Agricultural Outlook 2011–2020*, OECD Publishing and FAO. Available online at: http://dx.doi.org/10.1787/agr_outlook-2011-en.
3 See note 2.
4 S. Tangermann, *Policy Solutions to Agricultural Market Volatility*, ICTSD Issue Paper No. 33/2011.
5 In reality, this is a very complex issue. In general, modern financial instruments are highly articulated and their understanding is based on extremely refined mathematical models.
6 Definition adopted by the CFCT (US Commodity Futures Trading Commission).
7 Today there are 15 main markets worldwide on which standardised derivatives are traded on agricultural commodities: 5 in the US, 2 in Europe, 5 in Asia, 1 in Oceania, 1 in Africa, 1 in South America.
8 One of the cases of manipulation of the financial markets in agricultural commodities is that of the Italian Ferruzzi group which in 1989 had purchased large volumes of futures on soya, not before removing from the market large quantities of this commodity first traded by the group itself.
9 Some hold that the expression refers to counter products sold by chemists for which no medical prescription is required and whose prices are not controlled. Others maintain that it refers to dealings that occurred in bars around the Chicago Stock Exchange once official dealings had stopped.
10 It was especially what is known in economic jargon as *supply shocks* which unsettled market equilibria. In particular, the world's main exporters adopted strategies of reducing reserves in order to bring about increases in international prices.
11 In October 1973 OPEC adopted an oil embargo in response to US support for Israel in the Yom Kippur war.

12 The EU initiative which eliminated customs dues and quantitative limits on imports from the 49 less developed countries.

Chapter 4

1 See IWMI – Country Policy Support Program report 11, *Water Supply, Water Demand and Agricultural Water Scarcity in China: A Basin Approach*. Available online at: http://www.iwmi.cgiar.org/assessment/files_new/research_projects/ICID_China_A_Basin%20Approach.pdf.
2 See B. Turner, *Islam Vol. II: State and Politics*, Routledge, London 2003.
3 C. Breisinger, O. Ecker, P. Al-Riffai, *Economics of the Arab Awakening: From Revolution to Transformation and Food Security*, IFPRI Policy Brief 18, May 2011. Available online at: http://www.ifpri.org/publication/economics-arab-awakening.
4 G. Belgrano, *Il triangolo dell'oppressione*, interview with Adel Jabbar, MISNA 8 March 2011 (In Italian). Available online at http://www.scuoladipolitica.it/static/magazine/Il-triangolo-dell-oppressione-462.aspx. The Iraqi-born sociologist uses the expression *triangle of oppression* to identify the common features of the Arab Spring as the reaction to political despotism, religious fanaticism and foreign interference.
5 M. Lagi, Z. K. Bertrand, Y. Bar Yam, *The Food Crises and Political Instability in North Africa and the Middle East*, NECSI, Cambridge USA, 10 August 2011. Available online at: http://necsi.edu/research/social/food_crises.pdf.
6 These were the data presented in March 2011 by the Vice President of the World Bank, Shamshad Akhtar, to the First Arab Development Symposium on Food and Water Security in the Arab World, in Kuwait. The keynote address may be consulted at: http://web.worldbank.org/WBSITE/EXTERNAL/COUNTRIES/MENAEXT/0,,print:Y~isCURL:Y~contentMDK:22864816~menuPK:247603~pagePK:2865106~piPK:2865128~theSitePK:256299,00.html.
7 See on this point the analysis made by Annia Ciezadlo, *Let Them Eat Bread*, Foreign Affairs, 23 March 2011. Available online at: http://www.foreignaffairs.com/articles/67672/annia-ciezadlo/let-them-eat-bread.
8 See note 3.
9 J. Byrne, Sugar supply gaps ongoing claim German confectioners, 6 June 2011. Available online at: http://www.confectionerynews.

com/Regulation-Safety/Sugar-supply-gaps-ongoing-claim-German-confectioners

10 Now CIAA is called FoodDrinkEurope. The press release on the G20 summit of June 2011 is available at: http://www.fooddrinkeurope.eu/news/press-release/g20-ministers-urged-to-seize-opportunity-to-agree-on-an-action-plan-to-addr/

11 B. McCarney, Can the world still feed itself?, *Wall Street Journal*, 3 September 2011. Available online at: http://online.wsj.com/article/SB10001424053111904787404576529912073080124.html.

12 In the framework of the Common Agricultural Policy, a common market organisation (CMO) is a set of measures and structures that make it possible to manage the market for a certain farm commodity on the part of the EU. Quotes from: The Point of View of Europe's Sugar Using Industry – European Parliament Hearing on Reform of the CMO for Sugar 13 July 2005.

13 For this product the recorded price increases are largely due to the political crisis which since December 2010 has hit Ivory Coast, which alone accounts for 25 per cent of the whole volume of cocoa exported worldwide.

14 See *In America scoppia la guerra dei chicchi*, Il Sole 24 Ore, 20 July 2011.

Chapter 5

1 J. Vidal, Ethiopia at centre of global farmland rush, *The Guardian*, 21 March 2011, with video. Available online at: http://www.guardian.co.uk/world/2011/mar/21/ethiopia-centre-global-farmland-rush.

2 S. Liberti, *Land Grabbing. Come il mercato delle terre crea il nuovo colonialismo*, Minimum fax, Rome 2011.

3 See http://www.karuturi.com/index.php?option=com_content&task=view&id=12&Itemid=211.

4 C. V. Roy, Punjab farmers to grow multiple crops in Ethiopia, *Business Standard*, 2 August 2011. Available online at: http://www.business-standard.com/india/news/punjab-farmers-to-grow-multiple-crops-in-ethiopia/444507/.

5 M. Davison, *Ethiopian Government Slashes Karuturi Global Land Concession by Two-Thirds*. Available online at: http://www.bloomberg.com/news/2011-05-04/ethiopian-government-slashes-karuturi-global-land-concession-by-two-thirds.html.

6 See M. Konthari, *Report of the Special Rapporteur on Adequate Housing as a Component of the Right to an Adequate Standard*

of Living, 2005. Available online at: http://www.ohchr.org/EN/Issues/Housing/Pages/AnnualReports.aspx.

7 K. Deininger, D. Byerlee *et al.*, *Rising Interest in Farmland – Can It Yield Sustainable and Equitable Benefits?*, World Bank 2011. Available online at: http://siteresources.worldbank.org/INTARD/Resources/ESW_Sept7_final_final.pdf.

8 L. Cotula, S. Vermeulen, R. Leonard, J. Keeley, *Land Grab or Development Opportunity? Agricultural Investment and International Land Deals in Africa*, IIED, FAO, IFAD, Rome 2009. Available online at: http://www.fao.org/docrep/011/ak241e/ak241e00.htm.

9 A table summarising all the estimates available on the size of the phenomenon is available in the report for the Committee on World Food Security of the United Nations: HLPE, *Land Tenure and International Investments in Agriculture. A Report by the High Level Panel of Experts on Food Security and Nutrition of the Committee on World Food Security*, FAO, Rome 2011. See also note 7.

10 Cited in F. Roiatti, *Il nuovo colonialismo – caccia alle terre coltivabili*, Egea, Milan 2010.

11 V. Aarts, *Unraveling the 'Land Grab' – How to Protect the Livelihoods of the Poor?*, Amsterdam 2009. Available online at: http://farmlandgrab.org/10653.

12 For a press review see: http://www.farmlandgrab.org and http://www.commercialpressuresonland.org. For a review of publications see Saturnino M. Borras Jr., Ruth Hall, Ian Scoones, Ben White and Wendy Wolford (2011): Towards a better understanding of global land grabbing: an editorial introduction, *Journal of Peasant Studies*, 38:2, 209–216. See also Saturnino M. Borras Jr., Jennifer Franco, From threat to opportunity? Problems with the idea of a 'code of conduct' for land-grabbing, *Yale Human Rights & Development Journal*, April 2010. Available online at: http://www.tni.org/sites/www.tni.org/files/Yale%20April%202010%20Borras_Franco%20CoC%20paper.pdf.

13 On this aspect see Food First Information and Action Network (FIAN). Available online at: http://www.fian.org/programs-and-campaigns/projects/land-grabbing. Amongst the institutional personalities, UN special rapporteur for the right to food, Olivier De Schutter, is one of the most active in promoting the analysis of the phenomenon in a context of human rights. See O. De Schutter, *The Green Rush: The Global Race for Farmland and the Rights of Land Users*, 52 Harvard International Law Journal 504 (2011). There is also a large quantity of documentation to be found on De Schutter's website: http://www.srfood.org.

14 See the site of the ILC http://www.commercialpressuresonland.org.

15 According to the OECD definition, foreign direct investment (FDI) seeks to put in place a 'long-lasting interest' of the investor in a foreign country. This type of involvement implies the existence of a 'long-term' relationship between the investor and the project as well as a 'significant degree of influence' in its management. In general, FDIs are encouraged as an integrating element in the global economy as they can raise the competitiveness of the companies taking part as well as improve technology transfers and competences between different economies (see OECD Benchmark Definition of Foreign Direct Investment). Foreign direct investors in agriculture, if well managed, can bring about the modernization of cultivation techniques, an increase in yield and the stabilization of food supplies in countries which host them (see UNCTAD – *Transnational Corporations, Agricultural Production and Development – World Investment Report 2009)*. FDIs are even more important if one thinks that to satisfy food requirements in 2050, annual gross investments totalling 209 billion USD will be required in developing countries, a sum that is impossible to reach on their own given their limited spending power. See FAO, *Foreign Direct Investment – Win–Win or Land Grab?*, Issue paper for the World Summit on Food Security, Rome, 16–18 November 2009. Available online at: ftp://ftp.fao.org/docrep/fao/meeting/018/k6358e.pdf.

16 See Von Braun, J., Meinzen-Dick, R., 'Land Grabbing by Foreign Investors in Developing Countries: Risks and Opportunities, Ifpri Policy Brief 13, April 2009, available online at http://www.ifpri.org/publication/land-grabbing-foreign-investors-developing-countries. See also the papers presented at the largest ever conference on the theme of 'Global Land Grabbing', in April 2011 at the University of Sussex. The proceedings are available online at: http://www.future-agricultures.org/papers-and-presentations/cat_view/1551-global-land-grab/1552-conference-papers?start=10. See also: When others are grabbing their land, *The Economist*, 5 May 2011. Available online at: http://www.economist.com/node/18648855. See HLPE, *Land Tenure and International Investments in Agriculture. A Report by the High Level Panel of Experts on Food Security and Nutrition of the Committee on World Food Security*, FAO, Rome 2011.

17 See *Blowing the Farmland Bubble*, Rabobank report, 12 July 2012.

18 See M. Kugelman, S. Levenstein, *LAND GRAB? The Race for the World's Farmland*, Woodrow Wilson International Center for Scholars, Washington DC, 2009. Available online at: http://

www.wilsoncenter.org/publication/land-grab-the-race-for-the-worlds-farmland.

19 P. De Castro, *L'agricoltura europea e le nuove sfide globali*, Donzelli, Rome 2010.

20 See *$53B Food Basket*. Available online at: http://www.zawya.com/story.cfm/sidZAWYA20110629051450/53B_Food_Basket.

21 See Center for Human Rights and Global Justice New York School of Law, *Foreign Land Deals and Human Rights: Case Studies on Agricultural and Biofuel Investment*, New York, 2010.

22 Source: Emergent Asset Management, in K. Hunt, *Africa Investment Sparks Land Grab Fear*, BBC News, 5 August 2009. Available online at: http://news.bbc.co.uk/2/hi/business/8150241.stm.

23 See Von Braun, J., Meinzen-Dick, R., 2009.

24 See Daewoo to cultivate Madagascar land for free, *Financial Times*, 19 November 2008, and Madagascar scraps Daewoo farm deal, *Financial Times*, 18 March 2009.

25 See note 8.

26 See T. Hilhorst, J. Nelen, T. Nata, *Agrarian Change under the Radar Screen – Rising Farmland Acquisitions by Domestic Investors in West Africa*, Papers and presentations at the International Conference on Global Land Grabbing, University of Sussex, 6–8 April 2011. Available online at: http://www.future-agricultures.org/index.php.

27 J. Vidal, C. Provost, US universities in Africa 'land grab', *The Guardian*, 8 June 2011. Available online at: http://www.guardian.co.uk/world/2011/jun/08/us-universities-africa-land-grab.Si tratta del rapporto *Understanding Land Investment Deals in Africa*, available online at: http://media.oaklandinstitute.org/special-investigation-understanding-land-investment-deals-africa.

28 Data collected using the World Bank Database. Homepage http://www.databank.worldbank.org/data/home.aspx. D. Dongo, *Land-grabbing: il caso dell'Etiopia. Mentre i cittadini muoiono di fame, il governo cede terreno alle imprese straniere*, Il Fatto Alimentare, 31 March 2011. Available online at: http://www.ilfattoalimentare.it/land-grabbing-il-caso-delletiopia.-mentre-i-cittadini-muoiono-di-fame-il-governo-cede-terreno-alle-imprese-straniere.html. S. Rodi, *C'è la carestia e l'Etiopia cede le sue terre*, Corriere.it, 3 August 2011. Available online at: http://www.corriere.it/ambiente/11_agosto_05/testo-etiopia_94739dca-bdef-11e0-99fd-c37f66002d24.shtml.

29 In spite of the large numbers of publications of various types that there are on the topic it is still impossible to map large-scale

acquisitions in a precise fashion. The summary that follows is based on the cross-comparison of various sources carried out in August 2011: report of intergovernmental agencies, NGOs, news agencies, journalists' articles and 'grey' literature.

30 See http://wikileaks.org/cable/2009/12/09ADDISABABA2900.html.

31 For a critical and documented approach on development aid see D. Moyo, *Dead Aid – Why aid is not working and how there is a better way for Africa*, Allen Lane, London 2009.

32 Central theme of the IFAD conference held 24 and 25 January 2011 in Rome. See http://www.ifad.org/events/agriculture/index.htm.

33 See Africa Research Institute, *Waiting for a Green Revolution*, Briefing Note 0902 November 2009. Available online at: http://www.africaresearchinstitute.org/files/briefing-notes/docs/Waiting-for-a-Green-Revolution-ASTIXSXA4A.pdf.

34 A. Bakr, *Pakistan Opens More Farmland to Foreigners*, Reuters, 17 May 2009. Available online at: http://in.reuters.com/article/2009/05/17/idINIndia-39682520090517.

35 F. Roiatti, *Il nuovo colonialismo – caccia alle terre coltivabili*, Egea, Milan 2010.

36 A. Diallo, G. Mushinzimana, *Foreign Direct Investment (FDI) in Land in Mali*, Deutsche Gesellschaft für Technische Zusammenarbeit (GTZ), Eschborn 2009. Available online at: http://www2.gtz.de/wbf/4tDx9kw63gma/gtz2010-0064en-foreign-direct-investment-mali.pdf.

37 See D. Doya Malingha, Karuturi plans 500 million USD investment in Tanzania food production, *Bloomberg*, 18 August 2011.Availableonlineat:http://www.bloomberg.com/news/2011-08-18/karuturi-global-plans-500-million-investment-in-tanzania-food-production.html. See C. V. Roy, Punjab farmers to grow multiple crops in Ethiopia, *Business Standard*, 2 August 2011. Available online at: http://www.business-standard.com/india/news/punjab-farmers-to-grow-multiple-crops-in-ethiopia/444507/.

38 For an overview of the commercial agreements and the model of Chinese penetration in Africa, see S. Gardelli, *L'Africa cinese*, Egea, Milan 2010.

39 *Chinese Investments in South American Agribusiness*, Rabobank Report, 28 July 2011.

40 P. Murphy, *Brazil Mulls Leasing Farmland to Foreigners*, 9 May 2011. Available online at: http://www.reuters.com/article/2011/05/09/us-brazil-land-idUSTRE74856K20110509.

41 V. M. Sathish, UAE eyes investment in Chinese agriculture sector, *Emirates 24/7*, 11 March 2011. Available online at: http://www.

emirates247.com/business/economy-finance/uae-eyes-investment-in-chinese-agriculture-sector-2.011-03-11-1.366765.

42 R. Callick, Chinese firms eye Aussie farmland, *The Australian*, 8 May 2008. Available online at: http://www.theaustralian.com.au/business-old/property/chinese-firms-eye-aussie-farmland/story-e6frg9gx-1111116309176.

43 The report *Understanding Land Investment Deals in Africa* can be downloaded from: http://media.oaklandinstitute.org/special-investigation-understanding-land-investment-deals-africa.

44 A carbon credit is a certificate for the emission of a ton of CO_2 or other equivalent greenhouse gas. Carrying out projects such as the production of non-fossil fuels which aim to mitigate emissions generate carbon credits that can be exchanged on the emissions market.

45 Cited in S. Lubove, *Being Like Soros in Buying Farmland Reaps Annual Gains of 16%*, 10 August 2011. Available online at: http://www.bloomberg.com/news/2011-08-10/being-like-soros-in-buying-farm-land-lets-investors-reap-16-annual-gains.html.

46 GRAIN, *Pension Funds: Key Players in the Global Farmland Grab*. Available online at: http://www.grain.org/article/entries/4287-pension-funds-key-players-in-the-global-farmland-grab.

47 Second Swedish National Pension Fund/AP2, *Annual Report 2010*. Available online at: http://www.ap2.se/Documents/ap2-ar2010-gb-web_final.pdf.

48 This is underlined in his speech to the Conference of the World Bank on land and poverty held in April 2011. Available online at: http://vimeo.com/23314644.

49 Quotes from the press release of Oakland Institute. Available online at: http://media.oaklandinstitute.org/press-release-understanding-land-investment-deals-africa.

50 B. Walker, *Claims of African 'Land Grab' Spark Controversy*, 12 June 2011. Available online at: http://articles.cnn.com/2011-06-11/world/africa.land.report_1_farmers-claims-traditional-lands?_s=PM:WORLD. The CEO of EAM Susan Payne also spoke to the Conference of the World Bank on land and poverty held in April 2011, available online at: http://vimeo.com/23314644.

51 With regard to the Burgess case, see F. Roiatti, *Il nuovo colonialismo – caccia alle terre coltivabili*, Egea, Milan 2010; the article by Jessica Silver-Greenberg *Land Rush in Africa*, 7 December 2009, available online at: http://www.businessweek.com/magazine/content/09_49/b4158038757158.htm; the row regarding the article by Calvin Burgess, owner of the Dominion

Farm, available online at: http://www.dominion-farms.com/response.pdf. Burgess also uses YouTube to put his point across: see http://www.youtube.com/watch?v=kuODE5zBFfg.

52 See J. Spoor, Dimensions of global landgrabbing, *The Broker*, 6 June 2011. Available online at: http://www.thebrokeronline.eu/Blogs/A-new-agriculture-for-food-security/Dimensions-of-global-land-grabbing. The declarations of Mr Borras were obtained by telephone interview on 9 August 2011.

53 See *From Conflict to Peacebuilding – The Role of Natural Resources and the Environment*, UNEP. Available online at: http://www.iisd.org/pdf/2009/conflict_peacebuilding.pdf.

54 See A. Nunow, *The Dynamics of Land Deals in the Tana Delta, Kenya*, papers and presentations, International Conference on Global Land Grabbing, University of Sussex, 6–8 April 2011. Available online at: http://www.future-agricultures.org/index.php. Mwangi S. Kimenyi and Njuguna S. Ndung'u, 2005, *Sporadic Ethnic Violence – Why Has Kenya Not Experienced a Full-Blown Civil War?*, Inter Region Economic, pp. 125 and 141. Available online at: http://www.irenkenya.com/reports/publications/ethnic%20violence%202005%20kimenyi%20ndungu.pdf. Peter Mwangi Kagwanja, 2003, Globalizing ethnicity, localizing citizenship: globalization, identity politics and violence in Kenya's Tana River region, *Africa Development*, 28(1 & 2): 140–141. Available online at: http://www.scribd.com/doc/6512154/Kagwanja-Globalizing-Ethnicity-Localizing-Citizenship. T. McVeigh, Biofuels land grab in Kenya's Tana Delta fuels talk of war, *The Guardian*, 2 July 2011. Available online at: http://www.guardian.co.uk/world/2011/jul/02/biofuels-land-grab-kenya-delta.

55 O. Visser, M. Spoor, 2011, Land grabbing in post-Soviet Eurasia: the world's largest agricultural land reserves at stake, *Journal of Peasant Studies*, 38(2): 299–323.

56 See the website of the Land Union of Ukraine: http://zsu.org.ua/index.php?option=com_content&view=category&layout=blog&id=67&Itemid=108.

57 Romanians fear foreigners' quest over local agricultural land, *Romania Business Insider*, 7 July 2011. Available online at: http://www.romania-insider.com/romanians-fear-foreigners-quest-over-local-agricultural-land/28076/.

58 Telephone interview held 1 August 2011.

59 The FAO Committee of World Food Security endorsed the Voluntary Guidelines on the Responsible Governance of Tenure of Land, Fisheries and Forests in the Context of National Food

Security on 11 May 2002. See http://www.fao.org/cfs/cfs-home/cfs-land-tenure/en/.

60 FAO, IFAD, UNCTAD, World Bank Group *Principles for Responsible Agricultural Investment that Respects Rights, Livelihoods and Resources – Extended Version*, 25 January 2010. Available online at: http://siteresources.worldbank.org/INTARD/214574-1111138388661/22453321/Principles_Extended.pdf. In the field of the regularization of investments IFPRI have identified five key elements for a code of conduct for the acquisition of land abroad, and six basic principles proposed by the German Ministry for Economic Cooperation and Development contained in *Development Policy Stance on the Topic of Land Grabbing*, Federal Ministry for Economic Cooperation and Development, Bonn 2009. Available online at: http://www.bmz.de/en/publications/type_of_publication/strategies/diskurs015en.pdf.

61 See S. Suàrez Monsalve, *Land: Not For Sale!*, in *Land Grabbing and Nutrition – Challenges for Global Governance*, Right to Food and Nutrition Watch, Stuttgart 2010, available online at: http://www.rtfn-watch.org/fileadmin/media/rtfn-watch.org/ENGLISH/pdf/Watch_2010/watch_engl_innen_final_a4.pdf; and O. De Schutter, *Large-scale Land Acquisitions and Leases: A Set of Core Principles and Measures to Address the Human Rights Challenge*, Office of the UN High Commissioner for Human Rights, 2009, available online at: http://www2.ohchr.org/english/issues/food/docs/BriefingNotelandgrab.pdf.

62 The communication of the Commission is available online at: http://ec.europa.eu/development/services/dev-policy-proposals_en.cfm/. The project report of the European Parliament is available online at: http://www.europarl.europa.eu/sides/getDoc.do?pubRef=-//EP//TEXT+REPORT+A7-2011-0284+0+DOC+XML+V0//EN.

63 S. Liberti, *Land Grabbing. Come il mercato delle terre crea il nuovo colonialismo*, Minimum fax, Rome 2011.

64 Interview held in Rome, at IFAD HQ, 18 August 2011.

Chapter 6

1 Action plan on Food Price Volatility and Agriculture. Meeting of the G20 Agriculture Ministers, Paris, 22–23 June 2011.

2 Such as insurance or social action to support incomes and the spending power of farming families.

3 See www.feedthefuture.gov.

4 Environmental public goods and services, by their very nature, have positive effects on society as a whole or on a given community through the adoption of strategies (such as a reduction in polluting behaviour, landscape conservation, hydro-geological protection) which are generally more complex and costly than those usually adopted.
5 The CGIAR, the Consultative Group on International Agricultural Research, is a strategic partnership involving countries, international organizations and private foundations that have the aim of supporting the work of various international research centres in the field of agriculture with the priority of combating hunger in the world.
6 Martin, W. and K. Anderson (2012), Export restrictions and price insulation during commodity price booms, *American Journal of Agricultural Economics*, 2012, vol. 94, issue 2, pages 422–427.
7 International Centre for Trade and Sustainable Development (ICTSD), *Policy Solutions to Agricultural Market Volatility: A Synthesis*, Issue Paper No. 33/2011.
8 In some cases the support for farmers translates into the practice of buying raw materials when the price is low and thus supporting the needier farmers; this is a type of withdrawal from the market of products which pushes prices up artificially, helping the more vulnerable members of the group at times of low demand (when the price is lower).
9 International Centre for Trade and Sustainable Development (ICTSD), *Global Food Stamps: An Idea Worth Considering?*, Issue Paper No. 36/2011. Available online at: http://ictsd.org/i/publications/111809.

INDEX